J. Barry Maynard

Geochemistry of Sedimentary Ore Deposits

With 149 Figures

Springer-Verlag
New York   Heidelberg   Berlin

J. BARRY MAYNARD
Department of Geology
H.N. Fisk Laboratory of Sedimentology
University of Cincinnati
Cincinnati, Ohio 45221 U.S.A.

Production: Richard Ruzycka

On the front cover: Pyrite-cemented breccia in shale-hosted
barite deposit of Arkansas.

Library of Congress Cataloging in Publication Data
Maynard, James B.
    Geochemistry of sedimentary ore deposits.
    1. Ore-deposits. 2. Rocks, Sedimentary. 3. Geochemistry. I. Title.
TN263.M39   1983   553.4   82-19462

Typeset by WorldComp, Inc., New York, New York.
Printed and bound by R.R. Donnelley & Sons, Harrisonburg, Virginia.
Printed in the United States of America.

9 8 7 6 5 4 3 2 1

ISBN 0-387-90783-1 Springer-Verlag New York Heidelberg Berlin
ISBN 3-540-90783-1 Springer-Verlag Berlin Heidelberg New York

*For my parents, in thanks
for their support and encouragement*

Preface

This book is an outgrowth of my interest in the chemistry of sedimentary rocks. In teaching geochemistry, I realized that the best examples for many chemical processes are drawn from the study of ore deposits. Consequently, we initiated a course at The University of Cincinnati entitled "Sedimentary Ore Deposits," which serves as the final quarter course for both our sedimentary petrology and our ore deposits sequence, and this book is based on that teaching experience. Because of my orientation, the treatment given is perhaps more sedimentological than is usually found in books on ore deposits, but I hope that this proves to be an advantage. It will also be obvious that I have drawn heavily on the ideas and techniques of Robert Garrels.

A number of people have helped with the creation of this book. I am especially grateful to my students and colleagues at Cincinnati and The Memorial University of Newfoundland for suffering through preliminary versions in my courses. I particularly thank Bill Jenks, Malcolm Annis, and Dave Strong.

For help with field work I thank A. Hallam, R. Hiscott, J. Hudson, R. Kepferle, P. O'Kita, A. Robertson, C. Stone, and R. Stevens. I am also deeply indebted to Bob Stevens for many hours of insightful discussion.

Many people have read and commented on portions of the book. These comments have sharpened the presentation considerably, especially where the reviewer and I disagreed. In addition to those mentioned above, I thank S. Awramik, A. Brown, M. Coleman, K. Eriksson, W. Galloway, M. Gole, D. Holland, K. Klein, J. Leventhal, B. Price, B. Simonsen, F. van Houten, J. Veizer, A. Walton.

The text was composed using WYLBUR, a program developed at Stanford University. The University of Cincinnati computer center provided invaluable financial and technical support, and I thank the editors at Springer-Verlag and the staff at WorldComp for their patience in dealing with this form of manuscript.

Cincinnati, Ohio
January 1983

J. Barry Maynard

Contents

CHAPTER 1.

Introduction

"I should see the garden far better," said Alice to herself, "if I could get to the top of that hill: and here's a path that leads straight to it—at least, no, it doesn't do *that*—" (after going a few yards along the path, and turning several sharp corners), "but I suppose it will at last. But how curiously it twists! It's more like a cork-screw than a path! Well *this* turn goes to the hill, I suppose— no, it doesn't! This goes straight back to the house! Well then, I'll try it the other way."

Alice in *Through the Looking Glass*
From Lewis Carroll, *The Complete Works of Lewis Carroll*. (New York, Vintage Books, 1976, page 156.)
Reprinted with permission of the publisher.

Ore deposits come in a bewildering range of types, and their study involves almost all subdisciplines of geology. Accordingly, it has been difficult to cover the subject matter of economic geology satisfactorily in one book or in one course. In practice, some sort of restriction is made. For North American geologists, there has been a tendency, now fading, to regard economic geology as synonymous with the study of hydrothermal deposits. At one time this was carried to the extreme of insisting that almost all deposits were, in fact, hydrothermal. More recently, the importance of sedimentary processes in ore genesis has begun to be appreciated, but the sedimentology necessary for the understanding of these processes is hard to integrate with the standard economic geology curriculum. It seems to me that a workable solution to this problem is to make divisions along the lines of the traditional areas of geology. Thus, one can approach volcanic ores from the viewpoint of a volcanologist, sedimentary ores from that of a sedimentologist. The ores are then treated within the context of the surrounding rocks, rather than as isolated entities. This book attempts to fill this purpose for sedimentary ores.

Most sedimentary ores are formed by chemical processes, and these processes are the subject of this book. Accordingly, I have largely excluded placers. Ores are defined to be economic accumulations of metals, and I have not discussed non-metallics. Most non-metallics have an evaporite origin, a style of deposition well covered elsewhere (e.g., Braitsch 1971). The remaining metallic deposits form a more or less coherent group whose precipitation is mostly controlled by Eh gradients. That is, a particular metal dissolves at, say, low Eh, then migrates along the gradient to be precipitated at high Eh, or vice-versa.

What, then, is meant by "sedimentary ores"? How do they differ from deposits classed as "syngenetic" or "strata-bound"? I use sedimentary in the sense of "formed by sedimentary processes." Almost all such ores are also strata-bound, but many strata-bound ores seem to have been formed by hydrothermal processes (see Chapter 7). Similarly, many sedimentary ores are syngenetic, that is, formed at the same time as the surrounding sediment. But there is also an important class of syngenetic ores that is volcanic in origin (see Chapter 8), and some sedimentary ores are epigenetic. It is useful to make the following distinctions (based on Tourtelot and Vine 1976):

> *Sedimentary*—Formed by sedimentary processes, at low temperatures and pressures near the Earth's surface, hosted by sedimentary rocks or soils.
>
> *Strata-bound*—Enclosed by essentially parallel layers of sedimentary rocks. May be cross-cutting within a layer.
>
> *Syngenetic*—Formed at the same time the sediment was deposited. Sedimentary examples are iron and manganese ores, volcanic examples are the Kuroko-type Cu-Pb-Zn ores.
>
> *Early diagenetic (or syndiagenetic)*—Formed after deposition of the sediment, but while still in contact with overlying water via diffusion through pore fluids. Pyrite in modern sediments obtains its sulfur in this way.
>
> *Late Diagenetic*—Formed after the sediment is closed to the overlying water, but from metals derived from within the same stratigraphic sequence. Some U deposits in tuffaceous rocks may have this origin (Chapter 6).
>
> *Epigenetic*—Formed after consolidation of the sediment and usually cutting across the strata; metals supplied from outside the system. Two varieties may be distinguished:
>
>> *Ground water-epigenetic*—Formed by cold surface waters. An example is roll-front U.
>>
>> *Hydrothermal-epigenetic*—Formed by hotter ($>100°C$) solutions, usually from depth. Carbonate-hosted Pb-Zn deposits are mostly of this type.

Two tools for the study of the geochemistry of ores that are used extensively in this book need some introduction: stable isotopes and equilibrium diagrams. The stable isotopes of S, C, and O are proving to be invaluable in the study of sediments, and I have tried to include the data on these isotopes that apply to sedimentary ores. For a general introduction to stable isotopes in geology, see Hoefs (1980); here I will briefly review some of the principles.

Stable isotope analyses are presented on a relative scale of so-called delta values. That is

$$\delta = 1000(R_{sample}/R_{standard} - 1),$$

where R is the ratio of the amount of the heavy to that of the light isotope. Thus, for sulfur,

$$\delta^{34}S/1000 = (^{34}S/^{32}S_{sample} / \,^{34}S/^{32}S_{standard}) - 1,$$

from which it can be seen that a sample with a ratio identical to the standard will have a delta value of 0 permil, a sample with more of the heavy isotope will have a positive delta, and one with more of the light, a negative delta. The terms light and negative or heavy and positive tend to be used interchangeably. For each element of interest, an arbitrary standard has been chosen. For S the choice was troilite of the Canyon Diablo meteorite, for O "standard mean ocean water," for C a fossil belemnite. Except for seawater, these primary standards are practically exhausted and measurements are made with secondary standards, but all results are reported in terms of the original standards.

In sedimentary systems, wide ranges of delta values are encountered, reflecting large fractionations. For S and C, biologic activity is the most important cause of fractionation; for O it is evaporation and precipitation of water. Fig. 1-1 shows that S in seawater, which is mostly in the form of SO_4^{2-}, has a narrow range of values around $+20$, but S in sedimentary rocks, which is mostly present in FeS_2, is much more variable and can be considerably more negative, a result of bacterial sulfate reduction. C (Fig. 1-2) shows a similar biological effect: marine carbonates are around 0 permil (as might be expected from the use of carbonate C from a marine fossil as a standard), but organic matter, both living and fossil, is appreciably lighter. Thus, carbonates whose carbon comes from oxidized organic matter can be detected by their light C. For O (Fig. 1-3), meteoric waters are noticeably lighter than seawater, which is caused by a preference for the lighter isotope during evaporation of water (see Hoefs 1980, fig. 31). O isotopes are also affected by temperature. As diagenesis proceeds and temperature in-

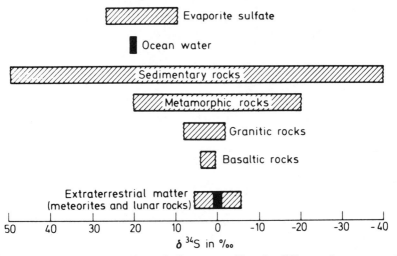

Fig. 1-1. Range of S isotopes in geologic systems. Note the difference between mantle-derived S and sedimentary sulfides (Hoefs 1980, fig. 12).

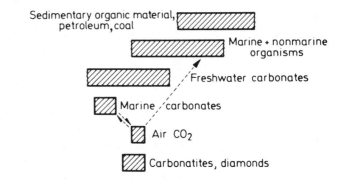

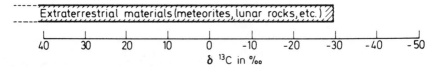

Fig. 1-2. C isotopes in geologic systems. Carbonate carbon derived from seawater is much heavier than organic carbon, and hence than carbonate formed by oxidation of organic matter (Hoefs 1980, fig. 9).

creases, O in minerals tends to become lighter. Thus it is vital to distinguish the timing of mineral formation when studying O isotopes.

The isotope ratios of these elements in seawater and sediments have changed with time, which has an effect on the values found in ore deposits. Changes in $\delta^{34}S$ of sulfate in evaporites, which is almost identical to that in coeval seawater, are particularly well documented (Holser and Kaplan 1966, Schidlowski and

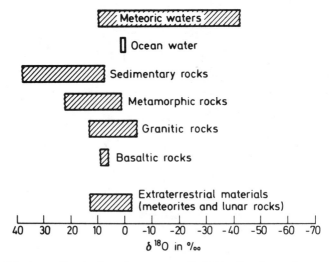

Fig. 1-3. O isotopes of meteoric water are generally lighter than those of seawater (Hoefs 1980, fig. 10).

Fig. 1-4.
S isotopes of evaporites and massive sulfide deposits show a parallel trend with time, suggesting that much of the S in the sulfides comes from seawater (after Sangster 1968, fig. 8).

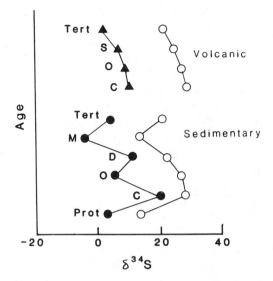

others 1977, Claypool and others 1980). Strata-bound sulfide deposits show a striking parallelism to this seawater curve (Fig. 1-4), providing strong evidence for the participation of seawater in their genesis. Veizer and Hoefs (1976) documented similar but smaller changes in C and O isotopes with time, and Veizer and others (1980) have shown that the S and C curves are correlated (Fig. 1-5),

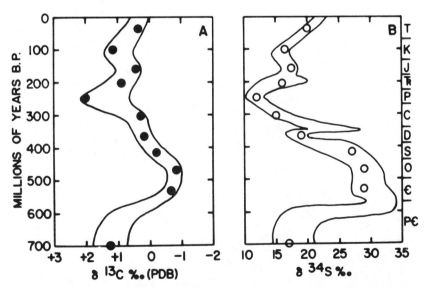

Fig. 1-5. Variation of C and S isotopes of carbonates and sulfates with time. Lower Paleozoic rocks have relatively light C and heavy S, indicating that a high proportion of the available C was deposited as carbonate, of the S as pyrite. Since the Carboniferous, the trend has been towards more organic C and sulfate S (after MacKenzie and Pigott 1981, fig. 3).

probably because of control of the reservoirs of reduced and oxidized S and C by tectonics, especially changes in sea level (Mackenzie and Pigott 1981).

Two shorter-term processes also can affect the S isotopes of an orebody. Schwarz and Burnie (1973) have presented a model in which S isotopes of sulfides in sedimentary accumulations are controlled by the degree of exchange between the water in the basin and the open sea. If exchange is rapid and complete, then S will be relatively light because bacterial reduction will only involve a portion of the total S with a strong preference for ^{32}S. If, on the other hand, exchange is limited, the bacteria will reduce all of the S coming into the basin, resulting in positive values close to coeval seawater. The other control is rate of sedimentation (Maynard 1980). If sedimentation is rapid, there is a concomitant increase in the rate of bacterial sulfate reduction. At high rates of reduction, the bacteria do not fractionate the S as efficiently, so the resulting sulfides are relatively positive (Fig. 1-6). The closed-basin effect seems to be important for the shales of the Kupferschiefer (Chapter 3), the sedimentation-rate effect for the uraniferous shales of the Appalachian Basin (Chapter 6).

Most geologists are familiar with equilibrium diagrams showing relationships between mineral phases and variables such as Eh, pH, temperature, or composition of the fluid. The standard treatment of this subject, which contains several applications to sedimentary ores, is Garrels and Christ (1965). Another more recent reference with a good short summary of these principles is Froese

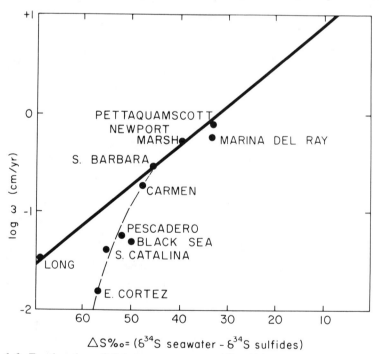

Fig. 1-6. Fractionation of S between seawater and sulfides increases as the rate of deposition (ω) decreases, up to a spread of approximately 50 permil (Maynard 1980, fig. 2).

(1981). Here I would like to make a few points about the use of such diagrams and sources of data for their construction. Any thermodynamic relationship can only apply when the system under investigation is at equilibrium, a situation notoriously rare in sediments. Why, therefore, are equilibrium diagrams used so often in this book? Such relationships tell us what *should* happen in a system, and thus provide a first approximation of its geochemical behavior. Then, by comparing this first model with observation, we can decide what other factors need to be considered. Usually these are reaction rates, an aspect of geochemistry that is, as yet, poorly understood. I have presented a few examples of kinetic studies for Mn and U, and have tried to identify other areas that could be studied in the same way.

Thermodynamic data for minerals are still in a state of flux, particularly for Al-bearing phases. I have, wherever possible, used the compilation of Robie and others (1978). Other sources, along with the values selected for use in the diagrams, are given in the Appendix. Please note that in this book the new international convention for expressing energy in joules (or kilojoules) rather than calories is followed. To convert, use the factor

$$1 \text{ joule } = 0.239 \text{ calories.}$$

Similarly, the equation relating the equilibrium constant for a reaction (K) to the change in free energy (ΔG_r),

$$\Delta G_r = -RT\ln K,$$

becomes, at 25°C,

$$\Delta G_r = -5.707\log K.$$

If the reaction involves oxidation and reduction, the standard electrode potential ($E°$) is also used:

$$\Delta G_r = 96.487 n E°$$

where n is the number of electrons, and ΔG_r is expressed in kilojoules. These relations can then be used with the standard equations for free energy change

$$\Delta G_r = \Delta G_f(\text{products}) - \Delta G_f(\text{reactants})$$

where ΔG_f represents the free energy of formation of the substance from the elements, and the Nernst equation

$$Eh = E° + (0.059/n)\log(a_{oxid}/a_{red})$$

where a_{oxid} and a_{red} are the activities of the oxidized and reduced species, to construct the type of diagrams used in this book. To use the Nernst equation in the above form, write the reactions so that the oxidized species are products. For instance,

$$Fe^{2+} \rightarrow Fe^{3+} + e^-.$$

Note that the electron should always appear on the right-hand side.

In cases where no thermodynamic data are available, there are several schemes

Table 1-1. Free energies of silication of various oxides to be used in calculating free energies of formation of layer silicates by the method of Tardy and Garrels (1974).

	Kcal	Kj
K_2O	-184.32	-771.21
Na_2O	-162.8	-681.2
MgO	-151.51	-633.9
CaO	-171.0	-715.5
FeO	-64.6	-270.3
Fe_2O_3	-177.50	-742.68
Al_2O_3	-383.2	-1603.3
SiO_2	-204.6	-856.1
H_2O	-57.8	-241.8
$Mg(OH)_2$	-201.7	-843.9
K_2O exchange	-184.32	-771.21

for making approximations. I have followed that proposed by Tardy and Garrels (1974), but see Lippmann (1977) for a critical treatment of these approximations. Basically, Tardy and Garrels' approach uses known values of free energy for sheet silcates to estimate the free energy of Al_2O_3, SiO_2, etc., within lattices of layer silicates, then uses these values to calculate free energies for unknown cases by simple summation. In order to use this approach in conjunction with the other free energy values in this book, it is necessary to correct their numbers slightly to account for different accepted values for their "known" minerals. To do so, use the numbers in Table 1-1. For example, a pure iron chamosite might have the composition $Fe_2Al_{1.43}Si_{1.43}O_5(OH)_4$. Its free energy of formation would then be estimated by first writing the formula as the sum of the oxides

$$2FeO + 0.72Al_2O_3 + 1.43SiO_2 + 2H_2O.$$

The free energy is then

$$2(-64.6) + 0.72(-383.2) + 1.43(-204.6) + 2(-57.8) = -813.28 \text{ Kcal,}$$
    or 3402.85 Kj.

If a mineral contains Mg, first assign as much as possible to $Mg(OH)_2$, then the remainder to MgO.

Such calculations can give only a crude approximation of free energy of formation of a mineral, but reasonable values usually result, keeping in mind the uncertainties in the "well-known" values. Further, the numbers so obtained can be used with some confidence in establishing the topology of equilibrium diagrams even if absolute values for the parameters plotted on the axes remain somewhat uncertain. Thus, the relationships among minerals can be explored, but there are limitations in comparing observed mineral assemblages with measured or inferred water chemistry.

CHAPTER 2.

Iron

"Would you tell me, please, which way I ought to go from here?"
"That depends a good deal on where you want to get to," said the
Cat. "I don't much care where—" said Alice. "Then it doesn't
matter which way you go," said the Cat. "—so long as I get
somewhere," Alice added as an explanation. "Oh, you're sure to
do that," said the Cat, "if you only walk long enough."

Alice and the Cheshire Cat in *Alice in Wonderland*
From Lewis Carroll, *The Complete Works of Lewis
Carroll*. (New York: Vintage Books, 1976, pages 71-72.)
Reprinted with permission of the publisher.

Iron has been mined from a variety of kinds of deposits, but production is now
almost entirely from two types: banded iron-formations and oolitic ironstones.
Banded iron-formations are rocks of mostly Precambrian age that are composed
of interlaminated quartz and iron minerals. They can be subdivided into two
varieties (Gross 1980): Algoma deposits, which are relatively small with an
obvious volcanic association, and Lake Superior, which are much larger and
have a shallow-shelf, orthoquartzite-carbonate association. The Algoma type is
abundant in the Archean, but an Ordovician example is found at Bathurst, New
Brunswick (see Chapter 8), and there is a possible analogue associated with the
Carboniferous Pb-Zn deposit at Tynagh, Ireland (see Chapter 7). Lake Superior-
type ores, by contrast, are confined to a particular time interval at around 2
billion years before the present. Oolitic ironstones have a more clastic association
than the iron-formations, and are higher in Al. Instead of banding, their most
prominent sedimentary structure is ooliths made up of hematite or chamosite.
Chert is rare, and they are found in rocks of a variety of ages from Proterozoic
to Pliocene.

As of 1970, these two types constituted more than 90 percent of the world's
iron production, with Lake Superior ores accounting for about 65 percent (Gross
1970, table 3). At that time, ironstones made up about 20 percent of the production, but this proportion has been decreasing: the last iron mines in Great
Britain, which are in Jurassic oolitic ironstones, were closing in 1980. The
extensive Lake Superior-type deposits of Australia are now coming into production, displacing smaller deposits, so that Lake Superior-type ores probably

make up 75 percent or more of production today. One reason for their predominance is a change in the nature of the feed for blast furnaces. Formerly, only ores that had been naturally enriched by the leaching of most of their SiO_2 (the "direct shipping ores") could be used. Since the introduction of pelletized feeds in 1955, there has been a shift to artificially concentrated cherty taconites because of lower shipping costs and lower energy costs for blast furnace operation (Kakela 1978).

Because of the differences in their chemistry and depositional setting, we will discuss iron-formations and ironstones separately. First, however, it is useful to examine some aspects of the distribution of iron in the Earth's crust.

Iron is the third most abundant metal, after Si and Al. Therefore, it is a major constituent of most rocks (Table 2-1), and its ores are actually rock types in their own right. The formation of these iron rocks depends on changes in oxidation state: Fe is mobilized under reducing conditions, precipitated under oxidizing. Thus, it will migrate from areas low in oxygen towards areas high in it, provided a pathway exists. Note, in Table 2-1, that igneous rocks, although varying in total iron, have about the same proportion of Fe^{3+} to Fe^{2+}. Sedimentary rocks, in contrast, show a wide range of oxidation states, indicative of the presence of environments of differing oxidation potential. Therefore, it is in sediments that the greatest potential for iron enrichment exists, and almost all iron ores are sedimentary. We will see that most of the other sedimentary ore metals also depend on redox gradients for their enrichment.

Table 2-1. Abundance of iron in common rock types.

	FeO (%)	Fe_2O_3 (%)	$\dfrac{Fe_2O_3}{FeO + Fe_2O_3}$
Igneous Rocks			
alkali-olivine basalts	7.9	4.2	0.35
tholeitic basalts	9.5	3.2	0.25
granodiorites	2.6	1.3	0.33
granites	1.5	0.8	0.35
Sedimentary Rocks			
Sandstones:			
quartz arenite	0.2	0.4	0.67
lithic arenite	1.4	3.8	0.73
graywacke	3.5	1.6	0.31
arkose	0.7	1.5	0.68
Shales and slates			
red	1.26	5.36	0.81
green	1.42	3.48	0.71
black	4.88	0.52	0.10
Seawater (as Fe)		0.007 ppm	

Source: Wedepohl 1969, tables 26-K-2, 26-K-3; Wedepohl 1971a, tables 6.2, 8.1, 12.2.

PART I. IRON-FORMATIONS

Almost every large area of Precambrian rocks contains iron-formation. Deposits of the Hamersley Basin of Australia, the Transvaal of South Africa, and the Lake Superior and Labrador Trough regions of North America are especially well described. There is no universally accepted terminology for these rocks; I have followed the usage proposed by Eichler (1976). The most general term is iron-formation (IF), which refers to a mostly Precambrian ferruginous-cherty rock. Banded iron-formation (BIF) is used for those varieties that show prominent banding of SiO_2 and iron minerals. Ironstone is used for the oolitic, non-cherty, aluminous iron ores more characteristic of the Phanerozoic.

A convenient classification scheme to use in discussing the mineralogy and geochemistry of these rocks is the facies system of James (1954). He separated IF into four types based on the predominant iron mineral: oxide (hematite or magnetite), carbonate (siderite or ankerite), silicate (greenalite, stilpnomelane, minnesotaite), and sulfide (mostly pyrite). In different localities the proportion of each facies varies, and the sulfide facies is poorly developed in Lake Superior-type ores. It should be realized that these are not facies in the normal sedimentary sense of a unified depositional system, because much of their mineralogy appears to develop during diagenesis (Dimroth and Chauvel 1973, Chauvel and Dimroth 1974, Dimroth 1977). In particular, the presence of the oxide facies is not sufficient proof that the rocks were deposited close to the margin of the basin (Walker 1978). However, inasmuch as the response of a sediment to diagenetic change is largely a function of primary sedimentary features, such as the amount of carbon, these facies are at least a reflection of original sedimentation. We will therefore use them as a framework.

Mineralogy

Oxides

The major problem in the mineralogy of iron-formations is defining the stage of the rocks' history at which various minerals developed, as illustrated by consideration of the origin of the two common oxides, hematite and magnetite. Dimroth and Chauvel (1973, p. 131) argued that the primary phases in IF were siderite, a "precursor silicate," amorphous SiO_2, and amorphous SiO_2 with adsorbed hydrated ferric oxides. Diagenetic processes involving reactions among these minerals and with externally added O_2, or with organic matter in the finer grained rocks, produced the magnetite, ankerite, and iron silicates now found. However, Klein (1973, 1974), using much the same petrographic data, maintained that magnetite and iron carbonates like ankerite and siderite were primary phases, formed by the recrystallization of fine-grained aggregates of the same mineralogy. These two pathways of magnetite formation are probably indistinguishable tex-

Table 2-2. Typical microprobe determinations of formulas for common minerals in IF. The Minnesota and Labrador deposits are Lake Superior type, the one from Western Australia is Algoma type.

	Fe	Mg	Al_{oct}	Al_{tet}	Si	O	OH	Source
Silicates								
greenalite								
Minnesota	4.87	0.56	0.06	—	4.24	10	8	1
Labrador	5.26	0.38	0.24	—	4.00	10	8	2
Western Australia	4.94	0.52	0.20	—	4.12	10	8	5
chamosite								
Minnesota	3.33	1.32	1.26	1.04	2.96	10	8	1
Labrador	4.00	1.35	0.85	1.25	2.75	10	8	3
Western Australia	3.62	1.11	1.28	1.36	2.64	10	8	5
stilpnomelane								
Minnesota	2.52	0.39	0.11	0.35	3.65	9.8	2.2	1
Labrador	2.16	0.52	0.16	0.25	3.75	10	2	2
Western Australia	2.36	0.30	0.01	0.38	3.62	9.0	3.0	5
minnesotaite								
Minnesota	1.40	1.60	—	—	4.00	10	2	1
Labrador	2.65	0.35	—	0.05	3.95	10	2	2
Western Australia	2.34	0.63	0.03	0.04	3.96	10	2	5

Table 2-2. (Continued)

	Fe	Mn	Mg	Ca	CO$_3$	
Carbonates						
siderite						
Minnesota	0.79	0.03	0.14	0.04	1.00	1
Labrador	0.70	0.06	0.21	0.03	1.00	4
Western Australia	0.81	0.05	0.12	0.02	1.00	5
ankerite						
Minnesota	0.52	0.06	0.34	1.08	2.00	1
Labrador	0.54	0.08	0.36	1.02	2.00	4
	Fe	As	S			
Sulfides						
pyrrhotite	47.3	—	52.7			5
arsenopyrite	33.7	30.2	36.1			5

Sources: 1. Floran and Papike (1975)
2. Klein (1974)
3. Klein and Fink (1976)
4. Klein (1978)
5. Gole (1980)

turally. They differ primarily in that Klein's implies that a closed system developed early in the diagenesis of the sediment, while Dimroth and Chauvel's requires migration of O_2 or of Fe^{2+} between beds. The scarcity of magnetite as an authigenic phase in modern sediments and unmetamorphosed younger sedimentary rocks suggests that either the magnetite in IF is, in fact, a late diagenetic product, formed at relatively high temperatures, or some condition was different in the Precambrian that favored the formation of magnetite during early diagenesis.

Carbonates and sulfides

Siderite is probably the most common carbonate mineral in iron-formation, followed by ankerite, although the difficulty in distinguishing these two in thin section may have led to a serious underestimation of the amount of ankerite. Most workers believe these carbonates formed at or near the sediment-water interface, but $\delta^{13}C$ values that have been reported are considerably more negative than in coeval calcite and dolomite, -10 permil compared with -2 to $+2$ permil (Becker and Clayton 1972, Perry and others 1973, p. 1122). Such light carbon, which is almost certainly derived from decomposition of organic matter, suggests that the Fe-bearing carbonates formed later, below the 'sediment-water interface, but probably before complete consolidation of the sediment. Microprobe analyses show considerable substitution of Mg and sometimes Mn in siderite (Table 2-2); dolomites and ankerites are Fe-rich, with a maximum of 0.65 mole percent Fe. Sulfides are seldom described from IF. Gole (1980) has presented two analyses from an Archean deposit in Australia (Table 2-2), but more data are needed for comparison.

Silicates

Previous work on the silicates has been reviewed by French (1973). The most common minerals are greenalite, a 7Å serpentine-like mineral; stilpnomelane, which has a talc-like structure, but able to accommodate considerable Fe^{3+} (Eggleton and Chappell 1978); and minnesotaite, an iron talc (Table 2-3). Structural details of these minerals can be found in Floran and Papike (1975). Chamosite, the common iron silicate of the Phanerozoic deposits, is sometimes found in shales associated with IF, but seldom in the IF itself. Nontronite, the common iron silicate from modern volcanic-sedimentary deposits (Chapter 8) is not found, presumably because it would, like other smectites, be destroyed by even moderate diagenesis. It is not known, however, to what mineral it would convert. Of the three most abundant minerals, greenalite and stilpnomelane are thought to be primary, and minnesotaite a product of very low-grade metamorphism (Ayres 1972, Klein and Gole 1981).

Almost ubiquitous in association with these iron minerals is some form of SiO_2, usually thinly interlaminated. The silica is now quartz or chalcedony, but was probably originally amorphous silica. Both SiO_2 varieties are referred to as chert, even if made up of macrocrystalline quartz. They have not been studied

Table 2-3. Common silicates in iron formation with hints on petrographic identification.

Mineral	Ideal Formula	Petrographic Characteristics
greenalite	$(Fe,Mg)_6Si_4O_{10}(OH)_8$	very fine-grained, nearly isotropic. Green, non-pleochroic.
stilpnomelane	$K_{0.6}(Mg,Fe^{2+}Fe^{3+})_6(Si,Al)_6(O,OH)_{22}$	mica-like form; pleochroic brown or green; moderate birefringence
minnesotaite	$Fe_3Si_4O_{10}(OH)_2$	fibrous; high birefringence, non-pleochroic

as much as the other constituents. Dimroth (1976, p. 218-220) reports that chalcedony (fibrous microquartz) can be either length-fast or length-slow, the latter variety being possibly associated with evaporites (Folk and Pittman 1971). The sequence of crystallization of silica minerals from amorphous silica has been studied extensively in recent years, both by experiment and by examination of Tertiary sediments (Kastner and others 1977, Murata and others 1977), and a promising field of future research is the application of these results to Precambrian cherts.

Metamorphism

Because so many iron-formations have been metamorphosed, it is useful for us to describe briefly some of the changes that occur. Furthermore, metamorphism is an important ore-forming process: it leads to an increase in grain size that makes gravity separation of the iron minerals from the quartz more attractive (Gross 1965, p. 55, 132). Iron-formation is not as sensitive to metamorphism as shaley rocks, particularly at low temperatures (Klein and Gole 1981), but three broad zones can be identified:

Low grade (200-350°C)—Defined by the first appearance of minnesotaite. More Na-rich rocks may produce crocidolite as well (Button 1976a, p. 279, Klein 1974, p. 486), although Australian workers (Trendall and Blockley 1970, Grubb 1971) have inferred an introduction of Na to form the large crocidolite and riebeckite deposits associated with the Hamersley IF.

Intermediate grade (350-550°C)—Characterized by the first appearance of grunerite-cummingtonite amphiboles, or if the fugacity of water is high, hydro-biotite (Immega and Klein 1976). This zone is also marked by the disappearance of most of the primary and diagenetic silicates and carbonates (French 1968, 1973).

High grade (>550°C)—Defined by the appearance of iron pyroxenes such as ferrohypersthene and hedenbergite (French 1973, p. 1066). Occurrences of rocks of this grade have been described from Montana (Immega and Klein 1976) and the southern end of the Labrador trough (Klein 1966, Kranck 1961).

Naturally the fugacity of water, O_2, and CO_2, as well as temperature, have an effect on metamorphism of these rocks. See Frost (1979) for a discussion. The approximate relationship of these zones to those in pelitic schists are shown in Table 2-4.

Supergene Enrichment

The other important ore-forming process is supergene enrichment. In this process, both leaching of silica to produce relative enrichment and downwards transport of iron leading to absolute enrichment are important. Most workers seem to agree that the soft, oxidized, "direct shipping ores" are produced in this way (Symons 1967, Porath 1967, MacDonald and Grubb 1971), but hard hematite orebodies also occur whose origin is more controversial. Some workers believe that these are formed by hydrothermal solutions (Bailey and Tyler 1960, Dorr 1964, Berge and others 1977), others regard them as metamorphosed supergene deposits (Cannon 1976), while others regard them as having been produced by normal weathering (Park 1959).

Detailed work by Morris (1980) on the iron ores of Western Australia suggests that most, if not all, of the features observed can be explained by weathering. He identified a sequence of alterations which he placed on a scale of increasing maturity. The least altered deposits contain martite (hematite pseudomorphous after magnetite) and goethite. Primary hematite remains unaltered while chert, carbonates, and iron silicates are leached out or replaced by goethite. In these ores, iron content averages 62 percent with a loss on ignition of 4 to 6 percent. At the opposite extreme, the most altered ores contain abundant secondary hematite with platy texture. This hematite appears to have formed via dehydration of the goethite formed in earlier stages. Intermediate types are found in which both fine-grained goethite and the coarse platy hematite occur. The formation of the platy hematite results in an increase in permeability, and thus in more complete leaching. Total iron is 64 percent, and loss on ignition is reduced to 1 percent. Further, phosphorus, an undesirable constituent for steel manufacture, is reduced to 0.05 percent, compared to 0.07 percent in the martite-goethite ores.

Geochemistry

In major element chemistry, Archean and Lake Superior-type iron-formations are virtually identical (Table 2-5), as pointed out by Gole and Klein (1981, fig. 3) and Gole (1981). The elements that show some difference are Al and P, which are higher in the Algoma-type deposits of Canada. However, the Al values reported by Gole and Klein (1981, table 1) for Algoma deposits are lower than

Table 2-4. Approximate correspondence of metamorphic grades in IF and agrillaceous rocks.

	Chlorite Zone	Biotite Zone	Garnet Zone	Staurolite Zone	Sillimanite Zone
Agrillaceous Rocks	chlorite ———	——— biotite ————————————————————————	garnet ———————————————————————	staurolite —————————————	sillimanite ———
Iron Formation	minnesotaite ———————————	hydrobiotite? ———————	grunerite – – – –	pyroxenes – – – – – – –	
Approximate diameter of typical quartz grains in chert layers (in mm)		.05	.10	.15	.20
Graphite Crystallinity (width of 002 peak at ½ height in °2θ)	1.61 to .62	.45	.38	.31	.22

Source: Adapted from French 1968, fig. 24, using data from Immega and Klein 1976. Graphite data are from Grew 1974.

Table 2-5. Bulk chemistry of Precambrian iron formations of Canada.

Major Elements (%)	SiO$_2$	Al$_2$O$_3$	Fe$_2$O$_3$	FeO	CaO	MgO	Na$_2$O	K$_2$O	P$_2$O$_5$	H$_2$O	S
All Facies:											
Algoma	48.9	3.70	24.9	13.3	1.87	2.00	0.43	0.52	0.23	1.4	1.57
Lake Superior	47.1	1.50	28.2	10.9	2.24	1.93	0.13	0.20	0.08	1.4	0.20
Oxide Facies:											
Algoma	50.5	3.00	26.9	13.0	1.51	1.53	0.31	0.58	0.21	1.1	0.29
Lake Superior	47.2	1.39	35.4	8.2	1.58	1.24	0.12	0.14	0.06	1.3	0.02
Silicate Facies:											
Algoma	46.2	7.56	15.8	18.1	0.83	3.89	0.05	0.41	0.42	3.5	1.85
Lake Superior	59.0	2.41	8.7	16.3	2.40	2.73	0.20	0.63	0.10	2.5	0.08
Carbonate Facies:											
Algoma	43.6	6.07	4.1	15.0	4.78	5.54	1.07	0.86	0.44	1.5	1.22
Lake Superior	38.1	1.40	5.1	21.2	5.12	4.54	0.15	0.15	0.15	1.5	1.13
Sulfide Facies:											
Algoma	42.5	6.23	15.1	14.6	2.27	2.42	0.91	0.73	0.17	3.0	10.73

Table 2-5. (Continued.)

Minor Elements (ppm)	B	Sc	Ti	V	Cr	Mn	Co	Ni	Cu	Zn	Ba
All Facies:											
Algoma	410	8	1240	109	118	1900	41	103	149	330	190
Lake Superior	210	18	390	42	112	4900	28	37	14	40	160
Oxide Facies:											
Algoma	160	8	860	97	78	1400	38	83	96	330	170
Lake Superior	240	—	160	30	122	4600	27	32	10	20	180
Silicate Facies:											
Algoma	—	11	1910	54	38	7300	36	151	99	340	70
Lake Superior	210	22	1900	124	102	3400	27	46	37	—	170
Carbonate Facies:											
Algoma	260	10	2570	140	357	4900	31	205	79	350	400
Lake Superior	150	16	330	68	79	7200	30	39	12	100	40
Sulfide Facies:											
Algoma	1800	10	2380	153	135	2000	57	115	466	290	140

Source: Gross 1980, table 3.

for the Canadian deposits, so this difference may be an accident of the sample used in compiling Table 2-5. Compared with ironstones (compare Table 2-10), these rocks have much lower Al and P and higher SiO_2.

In trace elements some differences do appear. Figure 2-1 shows that the two types of IF have parallel trends of composition compared with that of the Earth's crust, except for Mn, which is anomalously high in Lake Superior-type deposits, and Ni, Cu, and Zn, which are strongly depleted. These trends apply to the separate facies of IF as well as to the total (Table 2-5), except for the silicate facies, which has higher Mn in the Algoma instead of the Lake Superior-type. The apparent enrichment of the Algoma deposits in Al and in Ni, Cu, and Zn is consistent with their volcanic association and the common presence of massive sulfide deposits, as at Bathurst, New Brunswick. Algoma IF is also enriched in Au, having more than six times as much as in Lake Superior type deposits (Saager and others 1982a).

Some data are also available for rare earth elements (Table 2-6). Fryer (1977a,b) pointed out that Archean IF typically has a positive Eu anomaly, in contrast to the negative anomaly in modern ferruginous sediments and seawater. Proterozoic IF lacks a strong positive Eu anomaly but is still enriched in Eu relative to

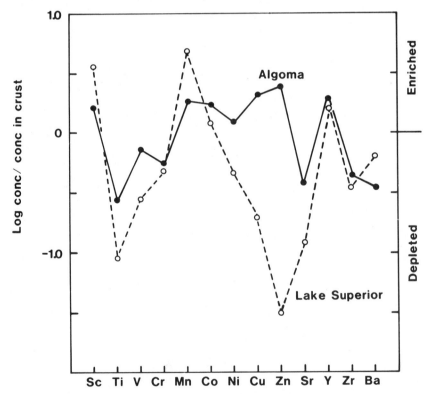

Fig. 2-1. Minor and trace elements in Canadian iron formations, normalized by crustal abundance of each element. (Data from Table 2-5.)

Table 2-6. Distribution of rare earth elements in iron-rich sediments (ppm).

	La	Ce	Nb	Sm	Eu	Tb	Yb	Lu	Source
Algoma IF									
Precambrian									
Michipicoten, carbonate	2.44	4.06	2.01	0.43	0.22	—	0.62	0.10	1
Soudan, pyritic slate	26	63	31	6.7	4.5	0.99	3.8	0.62	1
Mary River, hematite	2.55	3.92	1.97	0.39	0.21	—	0.29	0.058	1
Mary River, magnetite	3.93	6.17	3.60	0.60	0.39	—	0.66	0.11	1
Ordovician									
Bathurst, N.B., carbonate	16.8	13.9	—	3.08	4.19	0.9	1.38	0.21	2
Bathurst, magnetite	3.75	—	—	0.51	1.75	0.42	0.32	0.12	2
Bathurst, magnetite	4.55	—	—	0.82	2.54	—	0.54	0.13	2
Lake Superior IF									
Labrador, oxide	2.72	4.52	2.37	0.43	0.15	—	0.24	0.032	1
Labrador, silica-carb.	1.41	2.50	1.36	0.25	0.08	—	0.21	0.030	1
Australia, hematite	0.37	—	—	0.10	0.08	—	—	—	2
Australia, magnetite	2.60	—	—	0.32	0.11	—	0.82	0.19	2
Australia, stilpnomelane	25.8	50.3	—	6.84	1.65	1.65	5.89	0.85	2
Other Fe-rich Deposits									
East Pacific Ridge, flank	15.1	5.85	13.6	2.94	0.70	0.47	2.46	0.44	1
East Pacific Ridge, crest	28.7	18.0	28.6	5.62	1.34	0.95	3.14	0.53	1
Cyprus, ochre	66.7	21.7	59.0	14.3	3.60	—	3.92	—	1
Seawater $\times 10^6$	2.9	1.3	2.3	0.44	0.11	—	0.52	0.12	1

Source: 1. Fryer 1977b, table 2.
  2. Graf 1978, table 2.

contemporaneous shales. IF of both ages lacks the strong depletion in Ce found in modern seawater. He interpreted these differences as resulting from high concentrations of Eu^{2+} and Ce^{3+} in the Precambrian oceans, whereas in the present oxidizing oceans, Eu is present in the insoluble $3+$ oxidation state, and Ce^{3+} is oxidized to Ce^{4+} on manganese nodules. Thus, they are not available for incorporation into chemical sediments. The intermediate situation for the Proterozoic deposits would, then, reflect an intermediate average oxidation state for the oceans at that time. Graf (1978) disputed this interpretation, using as evidence the presence of an Archean-like pattern in the Ordovician age Algoma-type deposit at Bathurst, N.B. (Fig. 2-2). He contended that this pattern was not related to seawater chemistry but to the nature of water-rock reactions in seawater-hydrothermal systems. If so, one might expect a contrast in REE between Algoma and Lake Superior-type deposits, which is in fact indicated by the small difference in the Eu pattern (Fig. 2-2), although the two types are closely similar in the lighter REE from La to Sm.

As for the individual minerals, their non-stoichiometry (Table 2-2) makes them hard to model geochemically. For the oxides, carbonates and sulfides, end-member compositions can be used as a first approximation, but for the silicates we will calculate thermodynamic properties using the method of Tardy and Garrels (1974), as described in Chapter 1. The greatest uncertainty in the case of the

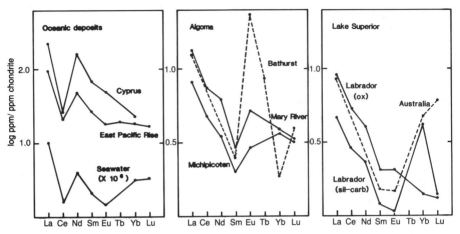

Fig. 2-2. REE in ocean deposits, Algoma IF and Lake Superior IF, normalized to chrondritic abundance. Each type seems to have characteristic Ce or Eu anomalies. (Data from Table 2-6.)

oxides and carbonates is in knowing which phases were present in the original sediment and which were formed during diagenesis. For the ferric oxides, we know that hematite does not precipitate from solution at low temperatures (Langmuir 1969, 1971). Instead, amorphous $Fe(OH)_3$ is the first phase to appear; if heated or aged it dehydrates to goethite, $FeOOH$, and eventually to hematite, Fe_2O_3. Thus, the proper phase to use in representations of the original sediment is $Fe(OH)_3$ rather than Fe_2O_3 (Klein and Bricker 1977). Likewise the first iron sulfide to appear is not pyrite, FeS_2, but mackinawite, FeS, which then converts to pyrite during diagenesis (Berner 1970). Because it is so prevalent in IF, one is tempted to include magnetite in this group of primary phases, as Klein and Bricker (1977) have in their series of diagrams, but the arguments just presented for hematite should apply to magnetite as well (Frost 1978). However, it is not known what the analogous amorphous precursor should be. Magnetite has been shown to form directly as a result of bacterial activity (Towe and Moench 1981), and such bacterial magnetite grains might have been an important constituent of Precambrian sediments; otherwise, there must have been some precursor, either $Fe(OH)_3$ that was subsequently reduced, or an amorphous mixed oxide. A possible candidate, which has been produced in the laboratory but not found in nature, is "green rust" (Feitnecht and Keller 1950, Schwertmann and Thalmann 1976), also referred to as "hydromagnetite" (Bernal and others 1959), which has a composition that can be represented as $Fe_3(OH)_8$. Feitnecht (1959) has shown that this substance will, in fact, convert to magnetite if allowed to age. Fig. 2-3 shows the resulting Eh-pH diagram, treating siderite as a primary phase because carbonates tend to crystallize readily without the kinetic problems of the oxides and sulfides. Note that the Fe^{3+} phase, which becomes hematite, and the Fe^{3+}/Fe^{2+} phase, which becomes magnetite, can coexist, as can siderite and the Fe^{3+}/Fe^{2+} phase, but the assemblage hematite-siderite should not occur, and it is, in fact, rare (Klein and Bricker 1977, p. 1467). Also, note that even at S

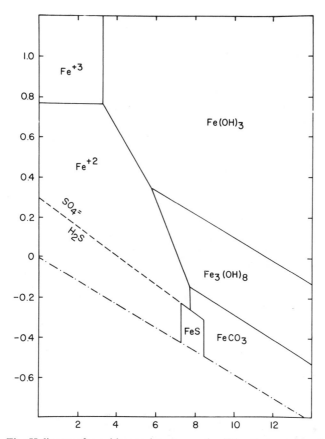

Fig. 2-3. Eh-pH diagram for oxides, carbonates, and sulfides that may have been present in the initial sediment of iron formations. 25°C, 1 atm; S,Fe = 10^{-6}.

concentrations as low as 10^{-6}, a considerable sulfide field is present. This is consistent with the presence of an important sulfide facies in association with Algoma IF, but the rarity of sulfides in Lake Superior IF indicates that S content must have been very low, less than 10^{-8} molar, compared with 10^{-2} in modern seawater. A similar diagram showing the stable phases in this system (Fig. 2-4) shows a vanishingly small region of stability for siderite, indicating that the use of the amorphous precursors in Fig. 2-3 is a better representation of the original sediment.

Next we need to add Si to this system. In the modern oceans, the concentration of silica is low because of silica-secreting organisms such as diatoms and radiolaria (Heath 1974). In the Precambrian, before the evolution of such organisms, the concentration must have been close to saturation with respect to amorphous silica or about $10^{-2.7}$ molar, and slight evaporation would have led to precipitation of amorphous silica. (Although this value is much greater than saturation with respect to quartz, none should form because quartz reactions are so slow that it is practically inert at surface temperatures.) Therefore, it is reasonable to suppose

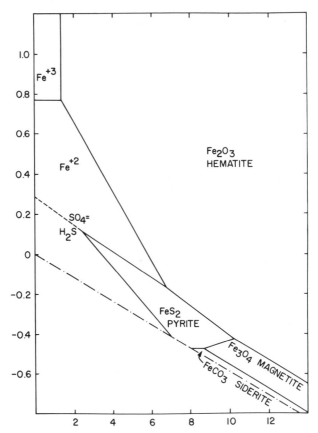

Fig. 2-4. Eh-pH relations for the eventual stable phases in the same system illustrated in Fig. 2-3.

that the chert in IF formed in this way. An alternative suggestion is that the precursor of IF chert was magadaiite, a sodium silicate with the formula $NaSi_7O_{13}(OH)_3$ that is found in certain tropical lakes (Eugster 1969, Eugster and Chou 1973). Thermodyanamic calculations suggest that at the pH and concentration of Na^+ of the modern oceans, magadiite should precipitate before chert during evaporation (Drever 1974, fig. 3). A problem is, what becomes of all of the Na^+? Some could be incorporated into riebeckite, but it seems to me unlikely that magadaiite could account for the volumes of chert found with no Na minerals. It would be interesting to study modern cherts formed via magadiite to see if there are any textural features that would be diagnostic of its former presence.

Besides chert, this Si must have also gone into iron silicates. The two most commonly cited as being primary are greenalite and stilpnomelane. If, however, we add stilpnomelane to our Eh-pH diagram (Fig. 2-5), it completely eliminates siderite, and nearly eliminates the magnetite precursor. The composition I used for this stilpnomelane, $K_{0.45}Fe_{6.2}Al_{0.85}Si_{8.15}O_{21}(OH)_6$, has only ferrous iron; if ferric iron is included, the field expands to cover even the magnetite field (Klein

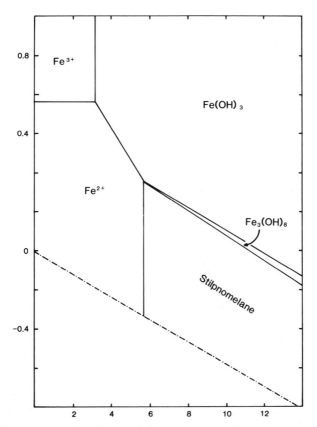

Fig. 2-5. Eh-pH diagram for IF with Si added, same conditions as for Fig. 2-3, but with $a_{H_4SiO_4} = 10^{-2.7}$.

and Bricker 1977, fig. 8). Greenalite is also theoretically unstable in the presence of stilpnomelane at the activity of H_4SiO_4 required to precipitate amorphous silica (Fig. 2-6). Thus, because chert-greenalite and siderite-greeenalite-chert assemblages are common, it seems unlikely that stilpnomelane can be a primary phase. In fact, Fig. 2-6 shows that greenalite most likely was not present as such, but as an amorphous precursor of the same composition. In these diagrams I have used a slightly aluminous composition for greenalite, $Fe_{5.5}Al_{0.20}Si_{4.1}O_{10}(OH)_8$, in accordance with the findings of Gole (1980) that this is the earliest silicate phase. The most likely primary assemblage was, then, amorphous greenalite, $Fe(OH)_3$, crystalline $FeCO_3$ and amorphous silica, possibly with an amorphous precursor of magnetite. During diagenesis these crystallized to greenalite, hematite, magnetite and chert, with stilpnomelane forming by reaction of one of the Fe-bearing phases with chert. Alternatively, the stilpnomelane could also have had a precursor phase, possibly nontronite. It would be interesting to see if nontronite could be converted to a stilpnomelane-like phase experimentally.

Studies of stable isotopes in IF have been confined mostly to carbon. For siderites, as already mentioned, $\delta^{13}C$ values tend to be lighter than in associated

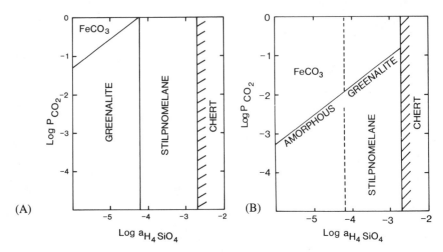

Fig. 2-6. A. Relationships among siderite, greenalite, and stilpnomelane in the presence of amorphous silica. Note that both siderite and greenalite should convert to stilpnomelane in the presence of chert. b. The same system, assuming that greenalite is amorphous and stilpnomelane is a later (diagenetic) replacement.

calcites and dolomites (Perry and others 1973, Becker and Clayton 1972), suggesting derivation of a part of the carbon from decomposing organic matter. Both of these studies were of Lake Superior-type deposits, however, and a single analysis from an Archean deposit gives $\delta^{13}C = +2.4$ permil (Goodwin and others 1976). It is possible that Algoma and Lake Superior-type deposits differ in the mode of siderite precipitation, but more study is obviously needed. Eichmann and Schidlowski (1975) have shown that the spread between carbon isotopes of organic matter and coexisting primary carbonates (limestone and dolomite) is the same as in younger sediments: -24.7 permil for the average organic carbon, $+0.9$ for carbonate carbon. Thus, the relative size of these two reservoirs of C, and the processes connecting them, must have been similar to those of today. C isotopes of limestone and dolomite associated with IF show little variation, but carbon in organic matter can sometimes be related to sedimentary facies. For example, Barghoorn and others (1977) found that in the Gunflint Chert of Ontario, the "algal chert" facies which is thought to be of shallow-water origin, contains organic carbon that is lighter (-25 to -30 permil) than that in the overlying, deeper-water chert-siderite and chert-magnetite facies (-15 to -20 permil).

The only study of S isotopes in IF is that of Goodwin and others (1976) on the Michipicoten Formation (Algoma-type). They found a wide range, from -10.5 to $+10.1$ permil, that is not related in any obvious way to rock type or stratigraphic position. Thus, it is not known whether the sedimentation rate or the restricted basin influences on S isotopes (Chapter 1) have any application here. This is an exciting area of study of IF, and a small investment of effort should produce much valuable information.

Oxygen isotopes of IF have been used to infer temperatures of metamorphism (James and Clayton 1962, Perry and others 1973, Becker and Clayton 1976, Perry and Ahmad 1981). For example, Becker and Clayton have shown that the Dales Gorge of the Hamersley in Australia was subjected to higher metamorphic temperatures than had previously been thought, 270-310°C. Further, they found that quartz, siderite and ankerite came to isotopic equilibrium, but that hematite underwent negligible exchange. Magnetite, on the other hand, seems to have come to equilibrium in some samples, but not in others. Hematite formation was confined to temperatures below 140°C, while most magnetite formation was at higher temperatures, suggesting that much of the magnetite in this IF is a late diagenetic product.

Petrography

IF shares many petrographic features with carbonate rocks, and some of the same terminology can be used. For example, Dimroth and Chauvel (1973) identified three principal textural components: matrix, cement and allochemical grains, as described in Table 2-7. They then defined seven textural types of IF based on

Table 2-7. Petrographic elements in IF.

Orthochems (precipitated at site of deposition)
    Fe micrite—microcrystalline siderite or iron silicate matrix.
    Matrix chert—microcrystalline quartz. May contain hematite dust in oxide facies IF; siderite and iron silicate inclusions in carbonate-silicate facies.
    Cements—fill pore space between allochems. Chert cements may be difficult to distinguish from matrix chert, but they lack the inclusions usually found in matrix chert.
    Quartz with columnar impingement (common).
    Chalcedony—fibrous quartz, oriented normal to allochem boundaries (rare).
    Fringing chalcedony followed by iron-rich sparry calcite.
    Fringing ankerite or minnesotaite followed by quartz.

Allochems (discrete, aggregated particles; probably transported)
    Pellets—ellipitcal bodies ~0.2 mm in long dimension found in matrix cherts. Composition is similar to matrix and so may be hard to distinguish. May have formed in place.
    Intraclasts—fragments of pre-existing sediment. In larger particles, internal sedimentary structures may be seen, but smaller varieties (<4 mm) are more problematic. Differ from pellets in being sorted, often cross-bedded.
    Ooliths and pisoliths—concentrically laminated particles, mostly composed of chert and hematite. Contain a variety of nuclei. The Sokoman IF has no Fe^{2+} minerals in ooliths.
    Shards—convex-concave bodies apparently derived from preexisting allochems by compaction.

Source: After Dimroth and Chauvel 1973.

Table 2-8. Textural rock types of the Sokoman IF.

Rock Type	Petrographic Elements	Ferric Types	Ferrous Types
Banded varieties			
Matrix chert	Matrix chert ± pellets	Laminated or ribboned matrix-chert plus hematite	Ribboned matrix chert with siderite, Fe^{2+} silicates or magnetite
Fe micrite	Fe micrite	Absent	Laminated Fe micrite ± magnetite
Varieties with abundant allochems			
Intra Fe micrite	Intraclasts + Fe micrite	Absent	Allochems with siderite, Fe^2 – silicate matrix
Intraclastic or oolitic matrix chert	Intraclasts + ooliths, pisoliths + matrix cement	Oolitic and intraclastic	Intraclastic
Cemented intraclastic	Intraclasts + ooliths, pisoliths + chert cement	Oolitic and intraclastic	Intraclastic
Recrystallized varieties			
Shard-bearing cherts	Shards + matrix or cement chert	Massive cherts ± patches of hematite, siderite, silicate	Massive cherts ± siderite, silicates
Recrystallized cherts	None remaining		

Source: After Dimroth and Chauvel 1973, table 2.

the proportion and nature of these components (Table 2-8). The major distinction is between the banded varieties, which must have been deposited in quiet water, and those with abundant allochems, which are commonly referred to as granular IF. Chauvel and Dimroth (1974, p. 321-322) infer that the granular type was deposited in shallow subtidal environments, the banded type in deeper, basinal water.

Textural features of the banded variety have been carefully described by Trendall and Blockley (1970, chapter 3) from the Hamersley basin of Australia. Chemical analyses have been presented by Ewers and Morris (1981), and Gole (1981) has described similar textures from Archean age IF. Unlike the Sokoman IF of the Labrador trough, which Chauvel and Dimroth (1974) studied, the Hamersley rocks are almost entirely of the banded type. There are three scales of banding. The first is a coarse *macrobanding* of alternating resistant IF and less resistant "shale." The IF bands range from 3 to 15 m averaging 6.4, while the "shale" bands are thinner, an average of 1.6 m. The shale macrobands differ from what are commonly thought of as shales in containing less detrital material and more iron minerals. They consist of a matrix of stilpnomelane or chlorite (chamosite?), often in random orientation, in which varying amounts of siderite, quartz, feldspar, and pyrite float. Laminations, where found, are defined by layers of pyrite or carbonaceous material. Shard-like structures are common, so much of the shale may be pyroclastic.

The IF macrobands are further subdivided into *mesobands*, which range in thickness from 1 to 80 mm, averaging 8 mm. Several varieties can be distinguished based on the predominant mineral (Table 2-9). Both macrobands and mesobands can be correlated over most of the outcrop area of the Hamersley, some 50,000 square kilometers (Trendall and Blockley 1970, p. 65-70).

The chert mesobands usually contain internal *microbands* defined by alternating iron-rich (magnetite, stilpnomelane, etc.) and iron-poor laminae, usually 0.5 to 1 mm in thickness. These microbands have been correlated over distances as great as 300 km, and are thought to be seasonal (Trendall and Blockley 1970,

Table 2-9. Mesoband types in the Dales Gorge Member of the Hammersley Basin.

Mesoband type (most abundant constituent)	Percent of section
Chert	
microbanded	46.5
non-microbanded	9.3
Chert-matrix	21.1
Magnetite	13.3
Stilpnomelane	2.3
Carbonate	1.0
miscellaneous	6.5

Chert-matrix differs from chert in having a uniformly very fine grain size. Only the chert mesobands show well-developed microbanding.

Source: After Trendall and Blockley 1970, table 4.

p. 106 and 257). Dimroth (1976, p. 204) has suggested that this microbanding is a diagenetic rather than a depositional feature, and, further, that these laminated IF types make up less of the section than the granular IF. The banded varieties, in fact, are overwhelmingly preponderant in most areas, and it seems unlikey that diagenetic processes could have preserved this delicate banding. The simpler hypothesis is to regard it as a primary depositional feature. The only other sedimentary rocks showing correlatable microbanding of this type are basinal evaporites (Anderson and Kirkland 1966). Accordingly, it is reasonable to regard the banded varieties of IF as forming in large, somewhat restricted basins by the evaporation of water enriched in silica and iron (Trendall 1973a and b, Button 1976a). IF displays a number of sedimentary structures. These include, in granular IF, cross-bedding (Gross 1972, Mengel 1973, Goodwin 1956), ripple-marks (Mengel 1973, Majumder and Chakraborty 1977), and various types of cut-and-fill structures (Gross 1972, Rai and others 1980). Markun and Randazzo (1980) have described in detail ooliths from the Gunflint IF of Ontario. Because of their similarity to calcareous ooliths and the absence of siliceous ooliths in modern marine environments, Markun and Randazzo (1980) interpreted them as replacements of pre-existing calcareous ooliths. Concretionary and dewatering features are seen in both banded and granular IF (Trendall and Blockley 1970, p. 153-162; Gross 1972; Majumder and Chakraborty 1977).

Grain sizes in granular IF have been reported by Mengel (1973) for deposits around Lake Superior. Mean size of the chert granules is ¼ to 1 mm (2 to 0Φ), with a symmetrical grain size distribution (low skewness) and moderate sorting (standard deviation = $0.5\ \Phi$ units). Some error is introduced in comparing thin section with sieve data, but using the equations of Harrell and Eriksson (1979) changes the range of average sizes by only 10 percent (2.2 to 0.2 Φ). Mengel (1973, p. 192) estimated the original density of the granules by comparing them with the size of detrital quartz in the same thin section. If they were deposited by the same currents, their densities are related by

$$\Phi_A - \Phi_B = 0.667(Log_2\rho_A - Log_2\rho_B)$$

where Φ is the phi mean grain size and ρ is the density of phases A and B. Solving for the density of the granules gives an original density of about 2.0, similar to that of opal. This result argues against an origin of IF by replacement of pre-existing carbonates; instead there seems to have been a primary, amorphous SiO_2 phase.

A distinctive aspect of the petrography of IF is its fossils. These are either micro-structural remains of bacteria or blue-green algae (Barghoorn and Tyler 1965, Barghoorn 1971, Awramik and Barghoorn 1977, Knoll and others 1978) or macroscopic features known as stromatolites (Hofman 1969, 1973, Awramik 1976). The morphology of stromatolites has been used to infer conditions in the environment of deposition such as water depth and current direction (Hoffman 1967, Truswell and Eriksson 1973, 1975).

Vertical Sequence

The next scale in our discussion of IF is an analysis of their vertical succession. For two of the best-exposed IF, in the Hamersley Basin of Australia and the Transvaal of South Africa, sequences have been compared in detail by Button (1976b). To these, I have added data from the Lake Superior region and the Labrador Trough (Fig. 2-7). An earlier comparison of the first three was made by Trendall (1968).

The tectonic evolution of the basins was similar. The IF-bearing sequence begins with a regional unconformity separating Lower Proterozoic sediments from Archean rocks. On this unconformity is a clastic unit, dominantly fining-upward, that was deposited in fluvial or shallow marine environments. In the Hamersley Basin, these clastics are accompanied by a great thickness of basic volcanics. This unit is followed by another unconformity except in the Hamersley. In the Lake Superior region and the Labrador Trough, a thick dolomite section, followed by a shale, intervenes below the unconformity. Next comes a transgression marked by a thin, clean sandstone. This is succeeded by a chemical unit composed of dolomite, IF, limestone, and carbonaceous shale in varying proportions and in various relative positions. In the Transvaal, the sequence starts with a thick dolomite that passes gradually into IF. A similar dolomite is present in the Hamersley, but is thinner and the section contains more shale. For the Lake Superior and Labrador deposits, dolomites are absent in this interval, and the IF lies directly on the basal sand in most places. This chemical unit is followed

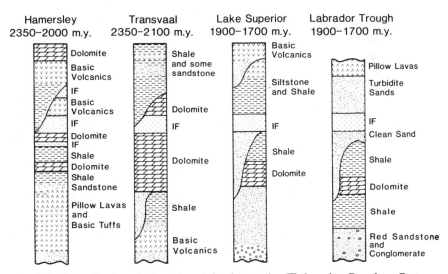

Fig. 2-7. Generalized geologic columns for four major IF deposits. Based on Button (1976), Beukes (1973, fig. 18), Trendall (1968), Bayley and James (1973, fig. 2), Rivers and Wardle (1979, fig. 4), and Gross (1968, p. 19-33).

by a regional unconformity, except in Labrador, and then by a thick clastic-volcanic sequence. Thus, the general sedimentary setting for Lake Superior-type deposits is shallow marine shelf with an orthoquartzite-carbonate association. The IF in the two younger deposits was formed early in a transgressive sequence, but, in the other two examples, a considerable thickness of other sediments intervenes between the basal sand and the IF.

What of the facies succession within the IF itself? One of the best-studied deposits from this standpoint is the Biwabik-Gunflint of Minnesota and Ontario. Fig. 2-8 shows two cycles, both characterized by a change from granular IF to laminated IF. The granular, or "cherty," IF contains abundant pellets of chert or chert plus iron minerals. The laminated, or "slaty," IF contains no pellets, instead it is characterized by a pronounced, fine banding, usually of chert and iron minerals, but sometimes including laminae of carbonaceous material. It has been interpreted as the deeper-water facies (White 1954). Accompanying the change from granular to laminated in the lower cycle is a progression in iron minerals from hematite to magnetite to carbonate-silicate (Fig. 2-8). Similar shallow-water granular and deeper-water laminated IF are found in the Labrador Trough (Chauvel and Dimroth 1974). Unlike these two, the Hamersley and Transvaal deposits have little or none of the granular variety, and thus are made

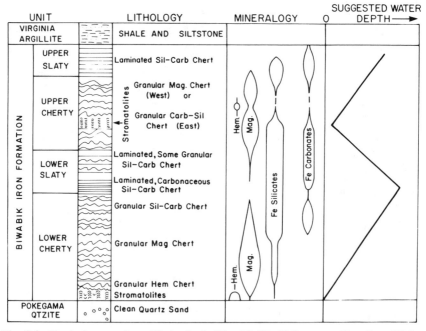

Fig. 2-8. Vertical succession of facies in the Biwabik IF of Minnesota. Data from White (1954, fig. 10), French (1968, fig. 2), Marsden and others (1968, table 4), Morey (1972, figs. 4-5 and 4-6), and Perry and others (1973, fig. 2).

up almost entirely of what we have interpreted as the deeper-water facies. It is
not known whether this difference is somehow related to the age of the deposits—
the Hamersley and Transvaal are older than the other two—or whether it is just
an accident of the small sample size.

In contrast to the Lake Superior type, the internal stratigraphy of the Algoma-
type IF is poorly known. One well-described example is the Michipicoten IF of
Ontario (Goodwin 1973a). Here, in contrast to Lake Superior deposits, most of
the associated rocks are volcanics or volcaniclastics. Also, there is a zonation
of mineral facies that Goodwin related to proximal-distal positions in the basin
(Fig. 2-9). Note the abundance of pyroclastics, the prominence of the sulfide
facies, and the absence of dolomite. The rocks appear to have formed in a
depositional setting similar to that of the volcanic-sedimentary ores in younger
rocks described in Chapter 8 (Fig. 2-10).

There is a possible third variety of IF characterized by an association with
sediments with glacial features. The best-known example is the Rapitan Group
of northern Canada (Young 1976, fig. 5), but others are known from South
America (Jacadigo Series), South Africa (Damara Supergroup) and Australia
(Umberatana Group). All are Upper Proterozoic, and so may be another peculiar
time-lithologic association like the Lake Superior deposits. As described by
Young, these deposits occur entirely within clastic strata of glacial or glacial-
marine origin. Also present are pseudomorphs after evaporite minerals. The IF
itself has not been described, but apparently consists of laminated hematite and
chert. It should be noted that this glacial-evaporite combination has caused
disagreement about whether the deposits are truly glacial, and if so, whether
they were formed at high latitudes (Schermerhorn 1974, Morris 1977, Williams
1978).

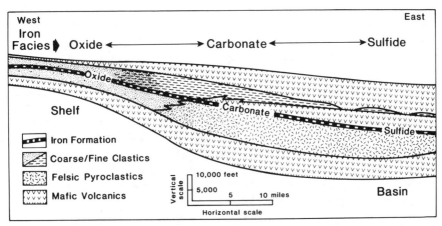

Fig. 2-9. Facies relations in the Algoma-type Michipicoten IF of Ontario according to
Goodwin (1973a, fig. 2).

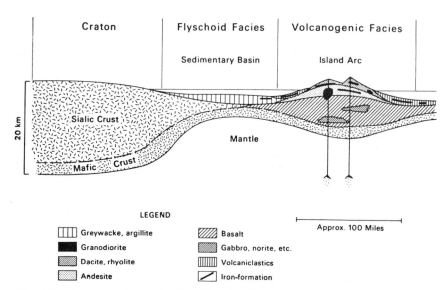

Fig. 2-10. Reconstruction of the depositional environment of Algoma-type IF (Goodwin 1973b, fig. 9). The setting is similar to that for Kuroko-type and other volcanic-sedimentary ores of felsic association.

Theories of Origin

Algoma-type IF are reasonably interpreted as being of sedimentary-volcanic origin. They are probably analogous to younger iron deposits formed by hydro-thermal circulation of seawater through volcanic rocks (Chapter 8). The Lake Superior-type deposits, on the other hand, have been most controversial. Their vast size and limited distribution in time are the two features hardest to explain. Most workers believe that the iron and silica were deposited directly on the sea floor, but a few (Cayeux 1922, Dimroth 1977), based on the similarity in sed-imentary structures between IF and limestones, have maintained that IF results from the replacement of earlier carbonates. The delicacy of the preserved textures and the difficulty of simultaneously introducing Si and Fe make this hypothesis unattractive. Furthermore, it does not solve the problem of how the large volumes of Fe and Si are derived.

Two sources of Fe commonly invoked for the Lake Superior-type deposits are rivers and volcanic emanations. The volcanic hypothesis is particularly at-tractive in the light of the precipitation of Fe in this way in modern sediments, and the similarity of Lake Superior to Algoma deposits in chemistry and min-eralogy. Gole and Klein (1981) have argued that these iron formations are in fact identical, and thus have a volcanic source. The question remains, however, as to why there should have been such an enormous outpouring of Fe between 2.4 and 1.8 billion years ago. Also, there are quantitative problems with both the volcanic and the river-supply models. Modern volcanic-hydrothermal systems do release Fe and Si, but the quantities are not large (Holland 1973, Maynard

1976). As a means of comparison, consider the Hamersley deposits. Holland (1973, p. 1169) has calculated that they contain 10^{20} g of iron that, if the microbands are varves as Trendall (1973c) suggested, were deposited in about 6 million years, or at a rate of 2×10^{13} g/yr. Ebeko, an unusually productive volcano in the Kurile Islands, releases 35-50 tons of Fe/day or a maximum of 1.8×10^{10} g/yr so that 1000 such volcanoes would be required to equal the rate attained in the Hamersley. This would be one for each mile of circumference of the basin, and nothing of the sort is suggested by the geology. If volcanoes were producing the Fe, there must have been a mechanism by which the site of precipitation was separated from the point of discharge of the hydrothermal fluids onto the sea floor.

The same kind of argument applies to supply by rivers. If there were no oxygen in the atmosphere at this time (Cloud 1972), iron could be carried in relatively large quantities by rivers as the more soluble $2+$ ion. The rise of oxygen in the atmosphere owing to the evolution of green-plant photosynthesis would, then, explain the disappearance of the deposits after about 1.7 b.y. Several workers (e.g., James 1954, Lepp and Goldich 1964) have inferred that such was the case, but the abundance of hematite in granular IF (Dimroth and Kimberley 1976), and the presence of red beds underlying the Sokoman IF of the Labrador Trough (Fig. 2-7), indicate that some O_2 must have been present in the atmosphere, at least during formation of the deposits of the Lake Superior region and the Labrador Trough (2.0 to 1.7 b.y.). In fact, it is reasonable to suppose, by analogy with Mars, that the Archean atmosphere had small amounts of free oxygen (Schopf 1981). Even in the absence of oxygen, only about 7 ppm Fe would be carried in solution by rivers, so that a river with 5 times the discharge of the Mississippi would be required to supply the Fe in the Hamersley (Holland 1973, p. 1170), and it is difficult to imagine what happened to the large clastic load such a river would carry. Again, some separation mechanism seems to be called for.

In 1952 Borchert proposed, for the Phanerozoic ironstones, the novel idea that the oceans were the source of the Fe (Borchert 1952, 1960). His ideas have not found wide acceptance because they require a structure and composition of the ocean that is not found today, and is unlikely for the Phanerozoic (see Hallam 1966, p. 315 and the discussion section of Borchert's 1960 paper). An oceanic source of the Fe in the Lake Superior-type deposits has been proposed by Holland (1973) and Drever (1974). Their model utilizes an intermediate level of oxygen in the atmosphere to provide a large reservoir of soluble Fe in the deep ocean at the same time that shallow, oxidizing environments existed where the Fe could be precipitated. Drever (1974, p. 1100) calculated that if atmospheric oxygen were 1/10 its present level, the oceans below the thermocline would be anaerobic, but still oxidizing at depths less than about 200 m (Fig. 2-11). The deep water in such an ocean would be enriched in Fe, either by remobilization of the ferric iron held on the surface of clastic particles (Carroll 1958) or by release of Fe^{2+} from hydrothermal systems at mid-ocean ridges. In the absence of silica-secreting organisms, this bottom water would also have been rich in dissolved silica.

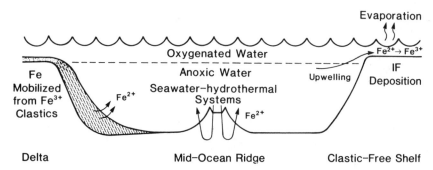

Fig. 2-11. Model for Lake Superior-type IF deposition (after Drever 1974). Deeper water, enriched in Fe^{2+} from either volcanic or diagenetic sources, moves up onto a shallow shelf and precipitates Fe and Si as a result of oxidation and evaporation.

Upwelling of this water onto shallow, clastic-free shelves would lead to deposition of Fe and Si by oxidation and evaporation. The situation is similar to that of the Bahama Banks today, where open ocean water, saturated with respect to $CaCO_3$, comes up onto the shelf and becomes supersaturated through warming and evaporation. Because the oxidation of Fe^{2+} leads to release of H^+, precipitation of $CaCO_3$ would be inhibited during periods of IF deposition; otherwise, these environments would be ideal for carbonate deposition, explaining the similarity in textures between IF and carbonates. Seasonal fluctuations in the rate of evaporation or the rate of supply of fresh seawater would explain the microbanding. The mineralogic facies observed in IF could be explained by an association of this upwelled water with increased production of organic matter, as is found in modern sediments (e.g., Brongersma-Sanders 1971). Then the minerals found would be a result of the amount of C originally deposited with the sediment (Fig. 2-12). Drever (1974, p. 1104) suggested that if there were an excess of Fe^{3+} over available C, magnetite resulted; if they were about equal, Fe^{2+} silicates formed; and if C greatly predominated, siderite was produced. Primary. hematite would have been preserved in shallow, well-oxygenated environments where all of the organic carbon was oxidized. The derivation of siderite from decomposing organic matter is supported by the isotopically light C found in siderite of Lake Superior-type deposits. Thus, this model explains the volume of iron, the absence of clastics and volcanics, the carbonate-like textures, the seasonal banding and the facies patterns. In addition, it explains the concentration of these deposits at a particular time in the Earth's history when atmospheric oxygen was intermediate between very low values in the Archean and essentially modern-day values in the Phanerozoic. The distribution of U ores also suggests a transition in O_2 in the atmosphere at about this time (Chapter 6).

There are, naturally, some difficulties. The first concerns S: why was all of the Fe not precipitated in deep water as iron sulfides, as it is in the Black Sea today, if the bottom waters were anaerobic? Drever (1974) gave two possible explanations. In the first, all of the S in deep water was precipitated by iron

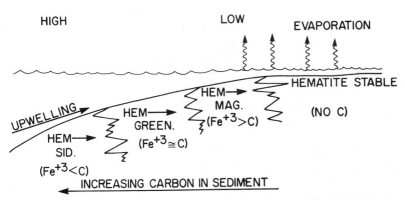

Fig. 2-12. A possible diagenetic explanation for the mineral facies of IF (after Drever 1974). Carbon content, controlled by productivity of algae and bacteria, is the ultimate control over mineralogy.

sulfides, but, because more Fe than S is being brought to the oceans by rivers, excess Fe^{2+} was available to be transported to shallow water. Note that this explanation requires there to be a low concentration of sulfate in the oceans and therefore little deposition of sulfate-bearing evaporites in the Lower Proterozoic. They are, in fact, rare, but it is not known whether this is a primary effect, or whether they have been preferentially removed by erosion. In a few places, casts of evaporite minerals have been found (Bell and Jackson 1974, Badham and Stanworth 1977), so at least some evaporite deposition was taking place. In the second explanation, the deep oceans had an oxidation potential low enough to convert Fe^{3+} to Fe^{2+}, but not so low as to allow the reduction of SO_4^{2-} to H_2S. However this requires a seemingly delicate balance of oxidation potential over the whole ocean. Another problem with the upwelling model is the low P content of these rocks. The environment described is one in which enrichment in P occurs today, but IF is notably deficient in it. Perhaps apatite as well as $CaCO_3$ was prevented from forming by lowered pH. Yet another problem is the absence of silicification of most Proterozoic carbonates. Why, if the oceans were so rich in dissolved silica, were all shallow-shelf sediments not enriched in SiO_2 throughout the Proterozoic?

The model outlined above explains most of the features observed in iron-formations, while introducing fewer problems than any of the other hypotheses that have been advanced. Note that it could be regarded as a variety of the volcanic hypothesis. That is, the iron and silica are ultimately of volcanic origin, but are transported to depositional sites remote from their site of introduction into seawater, thus accounting for the similarity in mineralogy and chemistry, but difference in sedimentary environment between Lake Superior and Algoma deposits.

Part II. Ironstones

The second common iron ore type is oolitic, such as the Jurassic Minette deposits of France and Great Britain, and the Silurian Clinton ores of North America. Such ores are common, although individual deposits are much smaller than the Precambrian ores. They are distinguished from iron-formation by a general absence of chert, oolitic instead of banded textures, and by mineralogy: their iron minerals are hematite or goethite, chamosite, and siderite, with no significant magnetite or low-Al iron silicates such as greenalite or minnesotaite. Bulk chemistry shows much higher aluminum, consistent with this difference in mineralogy (Table 2-10). Although the well-known deposits are from the Phanerozoic, there are Precambrian examples (Button 1976a).

These deposits have not generated as much interest as the iron-formations. Even though they are usually unmetamorphosed, and contain many fossil remains that can aid in deciphering their origin, there are fewer models for their formation, and their mineralogy and chemistry has not been as thoroughly investigated.

Mineralogy

The mineralogy of ironstones is comparatively simple. Oxides and hydroxides of iron, along with siderite and chamosite, make up almost all of the iron-bearing species. Pyrite may be present as finely divided particles or as replacements of ooliths or fossil fragments, but its undesirable metallurgical properties mean that pyrite-bearing ironstones cannot be used as ore (Taylor and others 1952, p. 456). Glauconite, the other common iron-aluminum silicate, is distinctive in its absence

Table 2-10. Bulk composition of some typical ironstones (%).

	Goethitic, Lorraine	Hematitic, Newfoundland	Chamositic, Northamptonshire	Sideritic, Yorkshire
SiO_2	5.4	9.9	13.4	7.8
Al_2O_3	4.3	3.2	7.4	8.4
Fe_2O_3	40.5	67.8	8.5	1.7
FeO	8.7	10.0	31.9	35.6
MnO	0.4	—	—	0.4
MgO	1.6	0.4	2.5	4.0
CaO	15.1	2.4	6.1	10.5
H_2O	7.2	2.8	8.5	4.2
TiO_2	0.2	0.4	—	0.2
P_2O_5	1.7	2.3	2.8	1.3
CO_2	15.0	1.1	17.4	25.3
S	0.04	—	0.5	0.1

Source: James 1966, tables 8, 9, 13, 16.

from ironstones, but it is common in associated rocks, and so will be discussed along with chamosite. Other minerals commonly found are kaolinite, calcite, phosphates, detrital quartz, and sometimes amorphous silica. The phosphates reported are low-iron varieties such as francolite rather than iron phosphates such as vivianite.

Oxides

Iron oxides. There are many oxides and hydroxides of iron known, but only hematite and goethite are commonly reported from ironstones. Some of the others probably occur and a knowledge of their distribution might be useful. For instance, maghemite has been produced in the laboratory under conditions close to those in nature (Schwertmann 1959) and so may occur in ironstones. There is a solid solution between maghemite and magnetite (Lindsley 1976, p. L22) and a mixed maghemite-magnetite mineral has been reported from calcareous sediments of the Indian Ocean (Harrison and Peterson 1965). Another possibility is β-FeOOH, which, in the laboratory, forms in preference to goethite in chloride-rich solutions such as seawater (Murray 1978). Some reported goethite or limonite may actually be this phase. There does not seem to be any information on the occurrence of the other iron hydroxide, lepidocrocite, in ironstones, nor on the relative distribution of the hydroxides and hematite.

Carbonates

Siderite (chalybite in older literature) is the only iron carbonate reported in significant amounts from ironstones. Ankerite is occasionally found (e.g., Smythe and Dunham 1947), but may commonly be overlooked. Unfortunately, the trigonal carbonates—siderite, dolomite, calcite, ankerite—are difficult to differentiate in thin section, but all may be important minerals in ironstones. One clue is that siderite has a higher index of refraction than the other common carbonates, 1.58-1.64; it is the only one with an index greater than Canada balsam (1.54). Also, the ferroan carbonates are often brownish as a result of slight oxidation, distinguishing them from low-iron varieties of calcite and dolomite. For further discrimination it is necessary to use X-ray diffraction or staining (Warne 1962). Careful application of these techniques, or microprobe analysis, to ironstone deposits may reveal minerals such as ankerite in unexpected amounts.

Sulfides

The common sulfides in sedimentary rocks are pyrite and marcasite. Only pyrite is reported from ironstones, but marcasite commonly occurs in similar sedimentary sequences (e.g., Maynard and Lauffenburger 1978), and so may have been overlooked in ironstone deposits. It is best differentiated from pyrite by using polished sections: marcasite has a distinct greenish bireflectance and a strong anisotropy, while pyrite is one of the few common isotropic sulfides.

The available laboratory (Rickard 1969a, 1969b) and field evidence (VanAndel and Postma 1954, p. 101) indicates that marcasite forms under lower pH conditions (<7), while pyrite forms at higher pH. If this proves to be the most important control on the distribution of these two minerals, their identification in sedimentary rocks would provide valuable evidence about conditions of deposition and diagenesis.

Silicates

The iron silicates of ironstones are considerably more aluminous than those found in iron-formations. Common examples are chamosite and glauconite; also reported are amorphous phases of similar composition and, occasionally, nontronite. The mineral name berthierine actually has precedence over the name chamosite (e.g. Brindley 1982), but chamosite seems to still be the preferred term, particularly in petrographic studies. Chamosite is a clay mineral whose chemistry is similar to iron-rich chlorites (in fact, in the French literature it is often referred to simply as chlorite) but which has a 1:1 sheet structure instead of 2:1:1 as in true chlorites. That is, it contains one silica and one alumina sheet in the unit structure instead of two silica, one alumina and one magnesium hydroxide; accordingly, it exhibits a 7Å instead of a 14Å periodicity in X-ray diffraction. Nelson and Roy (1958) termed such minerals septechlorites, which includes serpentines and greenalite as well as chamosite (Fig. 2-13). Most anal-

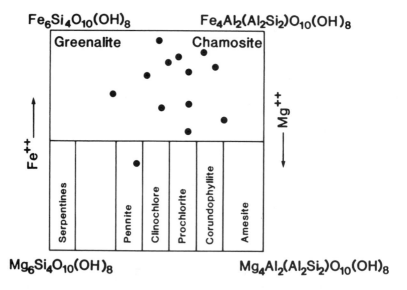

Fig. 2-13. Compositions of chlorite-like minerals (Nelson and Roy 1958). Dots are analyses of chamosites from Schoen (1964), Porrenga (1965), Weinberg (1973), Rohrlich (1974).

yses of chamosites show more silicon than the formula used in Fig. 2-13, $Fe_4Al_4Si_2O_{10}(OH)_8$. For instance, Schoen (1964) gave an average composition $Fe_{3.4}Mg_{0.8}Al_{3.0}Si_{2.8}O_{10}(OH)_8$.

True chlorite is also found in iron-rich forms such as thuringite or bavalite (James 1966, p. W7). These are usually associated with magnetite, suggesting that they form by very low-grade metamorphism of chamosite. Distinction of chamosite from chlorite in thin-section is difficult, but the chlorites tend to have a bluish-green pleochroism compared with green in chamosite.

Glauconite is also commonly present in iron-rich sedimentary rocks, although it is almost never found in the oolitic ironstones, and is itself never oolitic. Mineralogically, it is an iron-rich variety of illite-smectite. As in the aluminous illite-smectites, there is a close correlation between K content and the percent of non-expandable (illite) layers (Thompson and Hower 1975, fig. 2), although glauconite contains less K for a given proportion of non-expandable layers. Another difference between these two mineral series is that some glauconites have excess Fe^{2+} and Mg^{2+} over the amount that can be assigned to octahedral positions. Thompson and Hower (1975, p. 295) assigned these excess cations to interlayer hydroxy complexes; after subtracting these interlayer cations, the glauconite analyses match those expected for dioctahedral clays. Two examples are (Thompson and Hower 1975, table 8):

$$K_{0.73}(Al_{0.77}Fe^2_{0.41}Fe^3_{0.36} Mg_{0.52})(Si_{3.79}Al_{0.21})O_{10}(OH)_2$$

$$K_{0.36}(Al_{0.25}Fe^2_{0.39}Fe^3_{0.01} Mg_{0.41})(Si_{3.87}Al_{0.30})O_{10}(OH)_2.$$

Presumably, glauconites go through diagenetic changes similar to those undergone by aluminous illite-smectite (Hower and others 1976): during burial there should be an increase in K^+ accompanied by an increase in the porportion of non-expandable layers and an increase in the ordering of mixed layers.

Nontronite, Fe-smectite, has been reported from two localities. In modern sediments of Lake Chad, there are iron oxyhydroxide pellets that appear to be undergoing conversion to nontronite by the addition of silica (Pedro and others 1978). The reaction proceeds from the center of the grain outwards, producing a clay whose composition is approximately $X_{0.48}Fe_{1.87}Mg_{0.20}Al_{0.06}Si_{3.83} O_{10}(OH)_2$ where X refers to exchangeable cations such as Na^+ or Ca^{2+}. Note the low aluminum content. Petruk (1977) and Petruk and others (1977) have described nontronite and iron-rich opal from Cretaceous iron ores of western Canada (see also Mellon 1962). The nontronite in this case is much more aluminous, $Fe_{1.56}Fe_{0.31}Al_{0.99}Si_{3.45}O_{10}(OH)_2$. This deposit is perhaps unique among oolitic iron ores in containing large amounts of amorphous silica.

Both chamosite and glauconite are aluminous compared with the iron silicates of iron-formations. Why? Also, what factors dictate the predominance of chamosite over glauconite in ironstones? We shall approach these questions by considering the probable thermodynamic behavior of these phases and the available information on their distribution in modern sediments.

Geochemistry

The geochemistry of ironstones can be conveniently described by considering three problems: the relationship between siderite and pyrite, between glauconite and chamosite, and between the iron oxides and hydroxides.

Pyrite-siderite

Eh-pH relations among pyrite, siderite, and hematite (e.g., Berner 1971, fig. 10-2) indicate that pyrite is the only stable Fe^{2+} mineral at the sulfur level of modern seawater. Thus, siderite in ancient rocks suggests low-sulfur waters such as freshwater lakes and swamps (see Berner and others 1979, for a discussion of the relationship of iron minerals to salinity). Siderite is, in fact, common in fresh-water environments of modern deltas (e.g., Ho and Coleman 1969, p. 190), and a fresh-water origin has been proposed for siderites in coal measure shales of Great Britain (Curtis 1967).

On the other hand, siderite is commonly found in both shales and ironstones of undoubted marine origin. How does it form in these environments? Curtis has argued that, in marine shales, the rate of sedimentation is the controlling factor (Curtis and Spears 1968, p. 269; Curtis 1977, fig. 3). In this model, slow sedimentation permits complete conversion of reactive iron to pyrite, whereas more rapid rates shut off the sediment from contact with the the sulfur in seawater before all of the iron is converted to pyrite, thus allowing siderite to form. Several lines of evidence suggest that such a kinetic hypothesis is probably not correct. First of all, sulfide reduction, far from being suppressed by high rates of sedimentation, shows a steady increase with increasing sedimentation rate (Berner 1978, Maynard 1980). Thus, there is no reason to suppose it would stop before most of the available iron was consumed. Secondly, in most shales that I have studied, the transition from siderite- to pyrite-bearing is abrupt, suggesting a thermodynamic control, rather than gradual, as one would expect if a kinetic factor like rate of deposition were controlling. Finally, observations of electrode potentials in modern sediments show three distinct levels of oxidation potential, which probably correspond to predominance of the oxide, the carbonate, and then the sulfide (MacKenzie and Wollast 1977, fig. 2). It is even more difficult to see how sedimentation rate could control the appearance of siderite in ironstones; from all of the evidence it forms early in slowly deposited marine sediments.

The absence of siderite from Eh-pH diagrams for marine sediments can be overcome if, as was done in the previous section, primary phases such as $Fe(OH)_3$ and FeS are used instead of hematite and pyrite (Hem 1972, fig. 4). Such a choice is supported by the common appearance of $Fe(OH)_3$ and FeS as the first-formed iron compounds in sediments; during diagenesis they are transformed to the stable minerals (Berner 1971, chapter 10). With this set of phases, siderite does have a small stability field (Fig. 2-14) showing that its occurrence in marine sediments is consistent with thermodynamics and, thus, that kinetic arguments are not necessary.

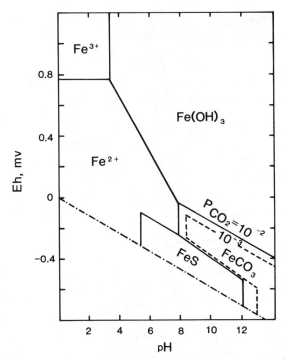

Fig. 2-14. Eh-pH diagram for some common phases in ironstone deposits. Siderite is a stable phase under normal seawater conditions (total S = $10^{-2.2}$) if $Fe(OH)_3$ and FeS are used instead of hematite and pyrite.

Note that this approach is the same as that originally developed by Krumbein and Garrels (1952) in their classification of sedimentary environments by Eh and pH. Berner (1981) has proposed a modification of this classification in which Eh and pH are not used as variables; instead environments are categorized by the presence or absence of oxygen and the presence or absence of dissolved sulfide. In this scheme, siderite is found in two diagenetic environments: first, where decomposition of organic matter has led to the depletion of oxygen in the sediment, but has not proceeded to the point of sulfate reduction, siderite forms in what he terms post-oxic environments; second, if so much organic matter is present that all of the available sulfate is consumed without using up the organic matter, methane fermentation will begin, and siderite will again be stable because so little dissolved sulfide is left. These two types of siderites should be isotopically distinct. In post-oxic environments, carbon in diagenetic carbonates is light, around − 14 to − 15 permil; in methanic environments, it tends to be much heavier, as much as + 15 permil (Irwin and others 1977).

The other mineral competing with siderite is calcite. For the reaction

$$FeCO_3 + Ca^{2+} \rightarrow CaCO_3 + Fe^{2+}$$
(siderite)   (calcite)

the data of Appendix 1 give a free energy of reaction of $+12.92$ kj. Thus, the equilibrium constant for the reaction would be

$$K = a_{Fe^{2+}}/a_{Ca^{2+}} = 0.0055;$$

siderite should be stable relative to calcite at Fe^{2+}/Ca^{2+} ratios higher than this value. For comparison, the data of Murray and others (1978) for anoxic waters of Saanich Inlet, British Columbia, show $a_{Fe^{2+}} = 80 \times 10^{-6}$ and $a_{Ca^{2+}} = 1.1 \times 10^{-3}$, after correction for complexing effects using the data of Davison (1979). The ratio is then 0.073, so siderite should be favored over calcite. Note that Berner (1971, p. 200) calculated a much higher equilibrium ratio, 0.05, for this reaction, using a different free energy value for siderite.

Chamosite-glauconite

Chamosite is virtually the only iron silicate encountered in ironstone deposits. Why are glauconite and greenalite so rare? Greenalite is probably absent because the sediments in which the ironstones formed were too rich in aluminum. Fig. 2-15 shows that at the very low aluminum concentrations found in seawater, greenalite should be stable relative to chamosite. Thus, chemical deposits such as iron-formation are greenalite-bearing. The aluminum to make the chamosite of ironstones was probably not supplied to the environment of deposition in a dissolved form, but rather by detrital clays, much in the same way that Carrol (1958) proposed that iron is transported into a basin. For instance, suppose that the aluminum to form chamosite is supplied by the dissolution of detrital kaolinite, as proposed by Schellmann (1969):

$$Al_2Si_2O_5(OH)_4 + 2.8Fe^{2+} + 3.6H_2O \rightarrow$$
(kaolinite)

$$1.4Fe_2Al_{1.43}Si_{1.43}O_5(OH)_4 + 5.6H^+.$$
(chamosite)

Then no aluminum or silica need be added to the system. Fig. 2-16 shows that the chemical compositions of chamosites are consistent with derivation from kaolinite by the addition of iron. Using this assumption, diagrams like Fig. 2-15 can be constructed.

For glauconite, the close similarity in structure to illite-smectite clays suggests that an analogous reaction can be used:

$$K_{0.68}Fe^2_{0.66}Fe^2_{0.66}Al_{0.94}Si_{3.8}O_{10}(OH)_2 + 3.22H_2O \rightarrow$$
(glauconite)

$$0.69K_{0.56}Fe^2_{0.49}Fe^3_{0.34}Al_{1.36}Si_{3.8}O_{10}(OH)_2 + 0.75Fe(OH)_3 + 0.29K^+ +$$
(illite-smectite)

$$1.17H_4SiO_4 + 0.14H^+ + 0.43e^-.$$

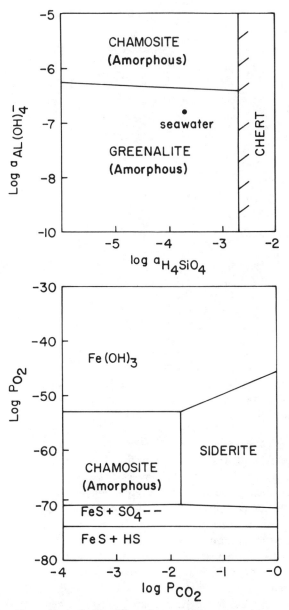

Fig. 2-15. *Top.* Greenalite is the Fe-silicate that would be expected to precipitate directly from seawater; chamosite requires an additional source of Al. *Bottom.* The amount of oxygen in the system determines whether iron precipitates as a sulfide, silicate, or oxide. Siderite occupies the same O_2 range as chamosite, but at higher CO_2. Equilibrium with kaolinite is assumed. pH = 8, S = $(10)^{-6}$.

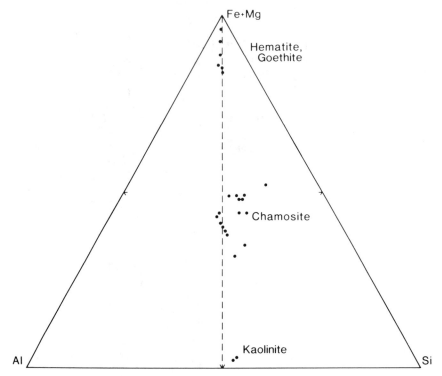

Fig. 2-16. Compositions of ooliths in ironstones of Newfoundland, Great Britain, France, and China as determined by electron microprobe.

From equations of this type, the relations shown in Fig. 2-17 can be calculated. I have inserted crystalline and amorphous fields for chamosite to show the range that can be expected. The youngest samples that I have examined are Jurassic, and these, although possessing an X-ray pattern, are so fine-grained as to be nearly isotropic and contain much more water than the ideal formula. Thus, the primary chamosite in these rocks is likely to have been cryptocrystalline, and to have occupied a position on the Eh-pH diagram intermediate to the two extremes shown. But, whatever the state of crystallinity of the chamosite, it can be seen to occupy a lower pH field than glauconite.

It is tempting to suppose that this difference in response to pH accounts for the distribution of these two minerals, but geologic evidence is against such an interpretation. Chamosite is known from only a few modern localities, but in all of these appears to form at normal marine pH, just like glauconite. Porrenga (1965, 1967) found chamosite in shallow waters of the Niger delta, glauconite in deeper water. He attributed this distribution to temperature, warmer water favoring the chamosite; but the subsequent discovery of chamosite in cold inland waters of Scotland (Rohrlich and others 1969) shows that this cannot be the case. Could it be that chamosite always forms in shallow water? Most modern and ancient occurrences suggest that this is an important factor, but Gygi (1981)

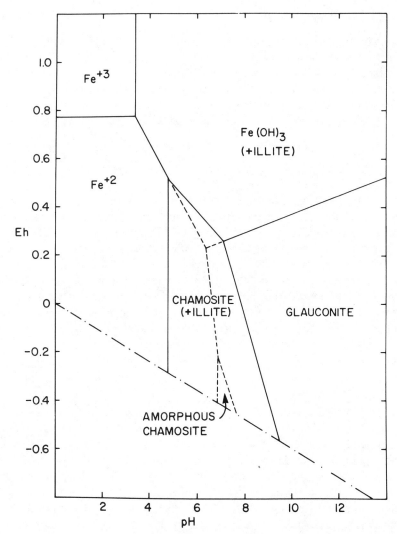

Fig. 2-17. Eh-pH relations between chamosite and glauconite, suggesting that chamosite should be favored by lower pH. Drawn for seawater activities of $K = 10^{-2.2}$ and $Al = 10^{-6.8}$; $Si = 10^{-2.7}$.

has presented thorough paleoecological documentation of a Jurassic ironstone from Switzerland that formed at a depth of about 100 m. Conceivably, the important factor is the mineralogy of the parent detrital clays. Most ancient occurrences have chamosite in the proximal, glauconite in the distal facies (e.g., Hunter 1970, p. 118). A similar trend is commonly found for kaolinite-illite distribution (Parham 1966, Edzwald and O'Melia 1975, Gibbs 1977), suggesting that there may be a primary depositional control over the distribution of iron silicates that is unrelated to diagenetic conditions. As yet the evidence for this idea is mostly circumstantial; more careful examination of the distribution of clay minerals in the shales associated with ironstones is needed.

Oxides and hydroxides

As already discussed, there is not much known about the details of the distribution of these minerals in ironstones. Hematite seems to be largely confined to the Paleozoic, the most common mineral in younger deposits being goethite (James 1966, p. W3). An exception is the Jurassic Minette ores of France, which I have observed to consist in a few cases of hematite, although goethite is more normal. Probably there is a progressive dehydration of the iron during diagenesis:

$$Fe(OH)_3 \rightarrow FeOOH \rightarrow Fe_2O_3.$$

Rarely, the final product of this sequence may be maghemite, γ-Fe_2O_3, rather than hematite, α-Fe_2O_3 (Deudon 1957).

Magnetite is sometimes found, and its presence poses some problems. In many cases it is probably the product of low-grade metamorphism, as in the Ordovician deposits of Brittany (Chauvel 1974) and Wales (Pulfrey 1933). In the Devonian of Libya, however, magnetite of apparent low-temperature origin has been reported (Hangari and others 1980). Here, oxygen isotopes of siderite and magnetite suggest temperatures as low as 40°C. It is not obvious what special circumstances favored the development of magnetite in this deposit; further, the oxygen isotope systematics of siderite are poorly understood. Thus, more work is needed on both the details of magnetite occurrences in ironstones and on oxygen isotopes before we can understand the behavior of magnetite.

Stable isotopes

Little work has been done on stable isotopes in ironstones, but a considerable body of information is accumulating on carbonate minerals in sedimentary rocks, much of which can be applied to ironstones. Carbon isotopes show a systematic variation with stage of diagenesis because of differences in the source of CO_2 (Irwin and others 1977). Four major sources are known, the first being seawater, which gives $\delta^{13}C$ values near 0 permil, while the others involve diagenesis of organic matter and yield distinctly different isotope ratios. In the sequence they appear during diagenesis, and with typical carbon isotopic composition of the CO_2 generated, these are

Oxidation:	$CH_2O + O_2 \rightarrow CO_2 + H_2O$	-20
Sulfate reduction:	$2CH_2O + SO_4^{2-} \rightarrow$	
	$2CO_2 + S^{2-} + 2H_2O$	-20
Methane fermentation:	$2CH_2O \rightarrow CH_4 + CO_2$	$+15$

Note that these reactions correspond to Berner's (1981) post-oxic, sulfidic, and methanic diagenetic environments. Because the CO_2 from these reactions mixes with that from other sources, a range of values is typically found. For instance, carbonates believed to have formed via oxidation or sulfate reduction typically have $\delta^{13}C$ values of -20 to -15 permil (Irwin and others 1977, table 1), presumably reflecting mixing with CO_2 from seawater. In the only study to date of siderite from ironstones, Hangari and others (1980) report $\delta^{13}C$ values ranging from -30 to -12, indicating a predominance of organic sources of CO_2.

Oxygen isotopes of siderite should be useful for determining the paleosalinity of the diagenetic environment: meteoric waters typically have a lower $\delta^{18}O$ than seawater. Unfortunately, neither the analytical fractionations nor the temperature effect are known with sufficient accuracy to permit unambiguous interpretations. In order to analyze carbonates for oxygen, they are usually dissolved in phosphoric acid, which, because it converts only two of the three oxygens in the carbonate to CO_2, produces a large analytical fractionation. For calcite, the appropriate fractionation factors are known, but they have not been investigated sufficiently in the case of siderite. To make matters worse, different workers have used different temperatures of reaction, making it impossible to compare results between laboratories. The same problem of temperature also applies to the diagenetic environment: the oxygen isotopic composition of a carbonate mineral is a function of both the temperature of the water from which it precipitates and its $\delta^{18}O$. Thus, either the temperature of formation or the $\delta^{18}O$ of the water must be determined by some independent method before the other can be determined by measuring oxygen isotopes of carbonate minerals.

With these limitations in mind, we can briefly examine some results for oxygen from siderite (Table 2-11). Timofeyeva and others (1976) suggested a salinity scale with values less than 20 permil indicating freshwater, heavier values brackish water or seawater. On this basis, Hangari and others (1980) interpreted Devonian ironstones of Libya, which have an average $\delta^{18}O$ of 19.7, to have undergone fresh-water diagenesis. Note, however, that the Canadian siderites

Table 2-11. Stable isotopes of siderite in shales and ironstones related to environment of deposition.

Environment	$\delta^{13}C$, mean	(range)	$\delta^{18}O$, mean	(range)
Pennsylvanian age shales, western Pennsylvania.[1]				
Freshwater, lam. gray	-4.75	$(-6.89/-3.62)$		
Brackish, lam., black	0.54	$(-1.38/3.91)$		
Marine, unlam., gray	1.53	$(-2.18/3.36)$		
Freshwater shales, Cretaceous of western Canada.[2]				
Edmonton Formation	10.6	(8.4/12.8)	22.9	(22.8/23.0)
Whitemud Formation	4.1	$(-1.1/7.7)$	23.3	(22.8/23.6)
Siderite nodules, Russian platform.[3]				
Freshwater			17	(10/18)
Brackish			24	(18/30)
Marine			26	(18/30)
Ironestones, Devonian of				
Libya.[4]	-18.4	$(-29.5/-12.1)$	19.7	(18.5/21.2)

Sources: 1. Weber and others 1964.
             2. Fritz and others 1971.
             3. Timofeyeva and others 1976.
             4. Hangari and others 1980.

investigated by Fritz and others (1971), although entirely freshwater in deposition and diagenesis, are much heavier than predicted, about 23 permil. Obviously much more work is needed.

Petrography

Ironstones are distinctive among iron ores in being predominantly oolitic. This texture is also the central enigma of their origin: how is it that a feature so characteristic of agitated, well-oxygenated water is found with ferrous minerals?

Careful descriptions of the petrography of the Jurassic ores of Great Britain have been presented by Taylor (1949, p. 17-23) and Dunham (1952, p. 16-33). Mineralogically, the ooliths consist of goethite, chamosite, or both. Siderite, or even apatite, may occur as occasional thin laminae, but siderite is more commonly seen as a matrix between the ooliths or as a replacement of pre-existing chamosite or goethite ooliths; no primary siderite ooliths seem to occur. The proportions of goethite and chamosite vary widely, even in the same bed, and some ooliths show evidence of oxidation of chamosite to goethite, particularly in surface exposures. In places, chamosite oolite grades into kaolinite oolite. The ooliths are usually set in a fine-grained matrix of chamosite or siderite; if a cement is present, it is usually sparry calcite. Also found in the section are non-oolitic beds of sideritic mudstone, sometimes rich in apatite.

The ooliths in ironstones are commonly flattened or irregular in shape. Knox (1970) separated chamositic ooliths from the Jurassic of Yorkshire into concentric and eccentric varieties. Both consist of alternating laminae of dark, randomly oriented and light, tangentially oriented flakes of chamosite. In the concentric variety there is a central nucleus, whereas the eccentric type has an off-center nucleus, and may have laminae on only one side of the grain. Both types tend to form as oblate spheroids rather than as spheres, owing to the development of an equatorial bulge. That is, growth was asymmetrical, resulting in an elliptical cross-section. This flattening is, then, a result of differing growth rates rather than deformation. Knox concluded that these ooliths did not form by accretion in agitated water, but in relatively quiet water from which they were subsequently washed up into oolite shoals. Similar mechanisms have been suggested by Mellon (1962, p. 938) and Chauvel (1974, p. 144). In contrast, Bhattacharyya and Kakimoto (1982) argued, based on the tangential orientation of the clay flakes, that chamositic ooliths do originate by accretion, but initially as kaolinite, which is converted to chamosite during diagenesis.

Chamositic and goethitic ooliths appear to differ somewhat in texture. For instance, James and VanHouten (1979, p. 129) reported that the goethitic ooliths from ironstone deposits of Colombia have a narrower size range, more varied nuclei, and are more spherical than the chamositic ooliths. Thus, it seems likely that in this case the two types of ooliths had a slightly different origin. Schellmann (1969) similarly inferred two origins, with chamositic ooliths formed in place in the sediment by reaction of dissolved iron with kaolinite, while goethitic ooliths formed in more agitated water where the iron was oxidized and the clay winnowed out. In this model the chamosite ooliths are a sort of micro-concretion while the goethitic ones are true ooliths. Many deposits, however, contain ooliths made up of alternating layers of goethite and chamosite (Alling 1947), presumably a reflection of oscillating chemical conditions, which suggests that both chamosite and goethite can be incorporated in ooliths by accretion.

Vertical Sequence

Ironstones occur almost exclusively in clastic sequences. Further, their position is usually close to the shoreline, even brackish in some cases; they also correspond in time to thicker sections in other parts of their basins of deposition. Three examples serve to illustrate these points: the Wabana ores of Newfoundland, the Clinton deposits of the eastern U.S., and the Jurassic ores of Great Britain.

The Wabana ores are early Ordovician in age and are confined to a small area of eastern Newfoundland. Slightly younger equivalents occur in North Wales (Pulfrey 1933) and in Brittany (Chauvel 1974) where the presence of magnetite and stilpnomelane suggests low-grade metamorphism. Ranger (1979) interpreted the Wabana deposits as having formed in a barrier-bar, tidal-flat complex based on sedimentary structures, trace fossils, and paleocurrents. The following description follows his work with some details drawn from Hayes (1915) and my own observations of the deposits. The sequence is entirely clastic; although trace fossils such as *Cruziana*, *Rusophycus*, and *Isopodichnus* are common, no calcareous shell material is present, only phosphatic brachiopods. Ranger (1979) divided the sediments into an offshore barrier-bar sequence and an inshore tidal-flat sequence. The barrier bars were usually composed of massive, clean quartz sands, but in two cases oolitic hematite substituted for the quartz, forming the Dominion and Scotia ore beds. The tidal flat sequence consisted of lower sand flats, middle mixed flats, and upper mud flats followed by a supratidal facies, perhaps analogous to modern salt marshes, which comprises thin, mudcracked black shales with interbeds of white, rippled sandstone. Just as with the barrier bars, the quartz of the lower sand flat facies is sometimes represented by oolitic hematite. Apparently, significant accumulations of iron were able to form whenever the supply of coarse detrital particles was reduced.

This nearshore complex developed adjacent to a delta which had a relief of at least 60 m based on the thickness of the prodelta foresets in one locality. The curious oolitic pyrite beds (Hayes 1915, p. 32) are found at the base of this prodelta sequence, just above one of the major oolitic hematite beds (Fig. 2-18). It is my belief that the pyrite formed during a relatively rapid transgression of anoxic basin waters across iron-rich shallow-water sediments. This transgression was then followed by a slower progradation of the delta back out into the basin, resulting eventually in the re-establishment of shallow-water conditions and the deposition of the second major hematite bed. A similar depositional pattern has been described by VanHouten and Karasek (1981) from Devonian

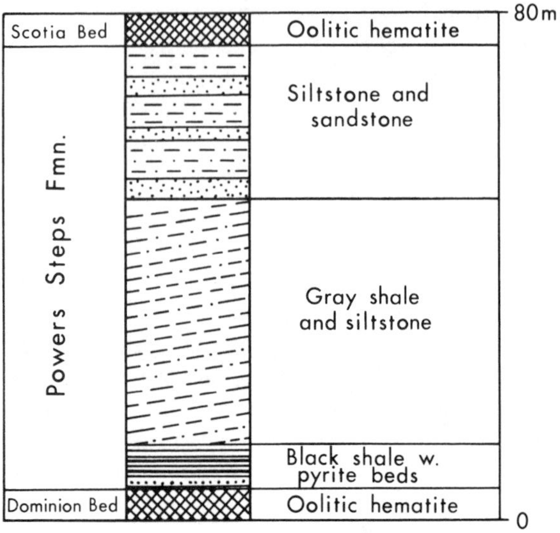

Fig. 2-18. Vertical sequences in the main ore-bearing section at Wabana, Newfoundland. The depositional cycle begins with transgression of reducing, basinal water over shallow-water clastics, followed by progradation of deltaic sediments, ended by deposition of oolitic ironstone as the last member of the regressive phase.

ironstones of Libya where oolitic chamosite-hematite beds accumulated at the top of coarsening-upward deltaic sequences during episodes of delta abandonment. Unlike the Wabana deposits, large vertical burrows penetrate the ironstones, and lag deposits of ferruginous phosphate overlie many of the ironstone beds.

Ranger (1979) theorized that the hematite ooliths formed originally as chamosite in lagoonal waters low in oxygen, and were subsequently washed up into the offshore bar or the lower sand flat facies, being oxidized to goethite (since dehydrated to hematite) in the process (Fig. 2-19). Microprobe analyses of the hematites (Fig. 2-16), however, show little aluminum, so derivation by oxidation of chamosite is unlikely, and the iron must have been precipitated directly as $Fe(OH)_3$.

The Silurian age Clinton ores of the eastern U.S. have stirred much interest, but the only regional sedimentological description is that of Hunter (1970). The ironstones are distributed over a wider range of facies than at Wabana, but are still predominantly found in what appear to have been lower intertidal to shallow subtidal areas. Hunter divided the Clinton group into eight facies types based on mineralogy (Fig. 2-20). Note that the facies distribution suggests deepening water to the west, until shallower water is reached again on the carbonate platform. Ironstones were best developed in shallow subtidal environments. Offshore bars, although present (Folk 1962), do not appear to have been important in localizing the ironstones. Each episode of ironstone deposition seems to be related to a general shallowing of the basin: ironstones in the distal part of the basin correlate with a tongue of sandstone prograding from the southeast (Hunter 1970, fig. 2)

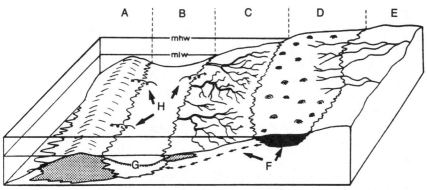

Fig. 2-19. Interpretation of paleoenvironments of the Wabana ironstone by Ranger (1979):
a. barrier bar; b. lagoon (low Eh sediments); c. tidal flat; d. supratidal algal
marsh (low Eh and high P_{CO2}; e. abandoned distributary. Iron was leached
from sediments of marsh environment (f), carried by groundwater to lagoon,
and precipitated as chamosite ooliths (g). These were subsequently reworked
into bar and tidal flat sediments (h).

and with an expansion of the carbonate facies. During Late Clinton Time, the
style of deposition changed to more open marine (Fig. 2-20); the *Lingula*-bearing
chlorite- and hematite-cemented sandstones were succeeded by carbonate-cemented
sandstones with a normal shelf fauna, and the green shales were succeeded by
more organic-rich gray to dark gray shales.

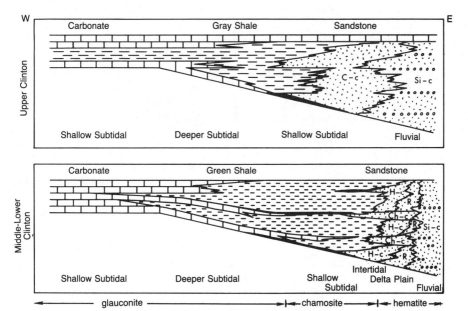

Fig. 2-20. Facies of iron deposition in the Silurian of the Appalachian Basin, with major
ironstone beds shown in black. Abbreviations: C-c, calcite-cemented sandstone; Si-c, silica-cemented SS; H-c, hematite-cemented SS; Ch-c, chlorite-
cemented SS; R, red, argillaceous SS (after Hunter 1970, fig. 2-4).

The best development of the Clinton ores is in the southern Appalachians near Birmingham, Alabama. Chowns and McKinney (1980) have presented a description of the facies succession that has many similarities to that at Wabana. Three coarsening-upward, progradational cycles can be recognized, each culminating in ironstone deposition. The cycle begins with prodelta turbidites and shales, which pass upward into hummocky-crossbedded sandstones with interbedded shales. This facies is interpreted as having been deposited below normal wave base, but in sufficiently shallow water to be affected by storms. Above these is a shallow subtidal to intertidal facies consisting of medium to coarse sandstone, coated with hematite and cemented with quartz. It seems to correspond to Hunter's (1970) silica-cemented sandstone facies, but the relationship of the other units to Hunter's mineralogic classification is uncertain. Cross-bedding in this facies is bipolar, as at Wabana, indicating tidal currents, and the trace fossil assemblage is mixed *Cruziana-Scolithus*. To the west, this facies passes into rich ironstone beds made up of abraded fossil debris in a hematite matrix. These were the self-fluxing ores of the Birmingham area. Oolitic ore beds are not common, but a few chamosite oolites and hematite-replaced chamosite oolites are known. Sheldon (1970) believed that, just as in Ranger's (1979) model for Wabana, the ore-grade beds were formed in lagoons (chamosite) or barrier bars (hematite). He further advocated an origin of the hematite by replacement of earlier-formed chamosite.

A number of ironstone deposits are found in the Jurassic of Europe. Their facies patterns, especially in Great Britain, have been described in a series of publications by Hallam (1963, 1966, 1967a, 1975, p. 39-46). Additional references to the continental European deposits can be found in Bubenicek (1964, 1968), Nicolini (1967), and in the series of papers in volume 31 (1979) of *Geologische Jahrbuch*. The ores are found predominantly in clastic associations, although some are highly calcareous (Taylor 1951, fig. 2), and they correspond in time to much thicker sequences elsewhere in the basin of deposition. Although brackish water conditions can be inferred for some deposits (Talbot 1974), most seem to have been deposited in water of normal marine salinity (e.g., Brookfield 1973). Rather than being near-shore or intertidal like the Wabana and most of the Clinton ores, they are thought to have been deposited on offshore swells (Fig. 2-21). This "Becken und Schwellen" style of deposition seems to have been more common in the geologic history of Europe than in North America, possibly accounting for the difference in the style of ironstone accumulation. The swells would have been responsible for the condensed section, and may also have protected the ironstones from dilution by clastics (the "clastic trap" hypothesis of Huber and Garrels 1953). Hallam (1967a, p. 431) suggested that the ironstones were an intermediate facies between clastics and carbonates, and that the type of sediment was controlled by the balance of two factors, the degree of water agitation, and the rate of terrigenous supply, which, in turn, was a function of proximity to rivers (Fig. 2-22).

In Yorkshire, where the section is almost entirely clastic, a vertical sequence much like that at Wabana is found. Hemingway (1951) described a sequence

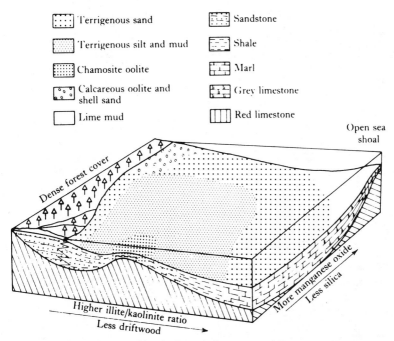

Fig. 2-21. Facies of iron deposition in Jurassic ironstones of Great Britain (Hallam 1975, fig. 3-9).

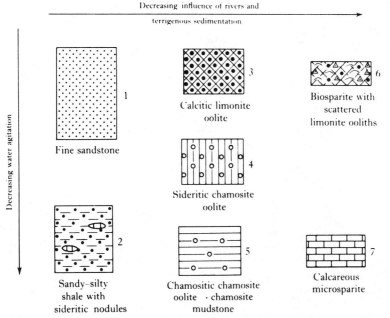

Fig. 2-22. Relationship of ironstone facies to current activity and rate of clastic supply (Hallam 1975, fig. 3-5).

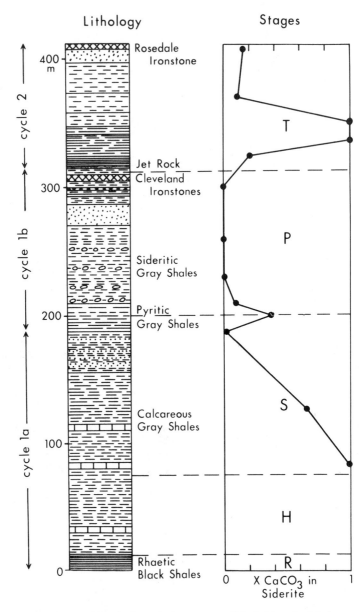

Fig. 2-23. Cyclic deposition in Jurassic ironstones of Yorkshire. In this interpretation, a cycle begins with deposition of a black shale, passes up through gray shales, then sandstones, and finally into oolitic ironstones. The lower cycle contains an incomplete progression in which no ironstone developed; instead renewed deposition of pyritic shale began. Letters refer to stage names: Rhaetic, Hettangian, Sinemurian, Pliensbachian, Toarcian. Compositions of concretions from Hallam (1967b).

black shale → gray shale → sandstone → ironstone that, based on analogy with the Black Sea, he interpreted as caused by shoaling of the water (Fig. 2-23). Because he believed the sideritic shales of the Lower Pliensbachian to be shallow-water ironstones, he identified three such cycles, but the rocks in question look more like normal prodelta shales with scattered siderite nodules. I have, therefore, divided the section into only two cycles, each beginning with a pyritic shale and culminating in ironstone deposition, with an incomplete cycle within the lower one representing Hemingway's first cycle. This interpretation is also more consistent with the data of Hallam (1967b), who found that the carbonate nodules in this section show a cyclic variation in the proportion of $CaCO_3$ to $FeCO_3$ (Fig. 2-23), which he related to proximal-distal variations, the pure $CaCO_3$ concretions forming in the most distal sediment.

Hallam and Bradshaw (1979) have used the concept of a progradational cycle of black shale and ironstone deposition to explain facies successions in the Jurassic and Lower Cretaceous. They proposed that the black shale at the base of the cycle formed during a rapid regional transgression. Thus, the base of each black shale is approximately a time line. Next, a coarsening-upward clastic sequence was deposited, culminating in shallow-water sandstones. After most clastic deposition ceased, the ironstone member of the cycle was deposited, followed by another transgression and renewed black shale deposition. It is not clear in this model whether a drop in sea level is necessary for the ironstone to form, or whether ironstone deposition followed submergence of the delta plain owing to continued compaction after migration of distributaries away from the site of deposition.

The above descriptions point to what we might call an ideal vertical sequence of ironstone deposition. It describes well the ironstones of four major areas: eastern Newfoundland, Alabama, Libya, and Yorkshire. A similar model has been proposed by van Houten and Bhattacharyya (1982), but with the addition of a burrowed, phosphatic hardground at the top of the sequence. How general this progradational model is needs to be tested by carefully examining the stratigraphic succession for other cases. For example, Late Jurassic ironstones of Switzerland were relatively deep-water deposits (Gygi 1981); further, some of the Silurian ironstones of the Appalachians are almost entirely in carbonates. Thus, not all ironstones formed in this model setting. More work is needed to find out how many did, what features of this depositional sequence particularly favored ironstone formation, and if those features can be found in some form in the anomalous cases.

Theories of Origin

No satisfactory general model for the origin of ironstones has been developed. Largely this failure is attributable to the absence of modern analogues, but, as indicated in the preceding discussion of the vertical sequence, ironstones may form in more than one way. The problem is best considered by separating it into two parts, the origin of the oolitic texture and the source of the iron.

Geologists have long been perplexed by the presence of ferrous minerals such as siderite and chamosite, which one would expect to form in quiet, low-oxygen water, in oolite, a rock normally associated with agitated, well-oxygenated conditions. In order to avoid this dilemma, Kimberley (1979) revived an old idea of Sorby and Cayeux that the ooliths were originally calcareous. He proposed that the ooliths were initially composed of aragonite that was subsequently transformed to chamosite and goethite by downward-percolating meteoric waters that leached iron from organic-rich fluvio-deltaic sediments immediately above the ironstone. The lack of carbonate sediments in most ironstone sequences, the evidence for early formation of the ooliths, and the rarity of the proposed stratigraphic succession have been cited as evidence against this model (James and VanHouten 1979, Hallam and Bradshaw 1979, and discussions of the original paper). Another problem is the aluminum content of these rocks. Oolitic limestones are not argillaceous, so that a considerable amount of aluminum, as well as iron, must be added. For chamosite oolite, the Al/Fe ratio is 0.40, for goethite 0.16 (James 1966, tables 8 and 13). Because aluminum is one of the least mobile elements in sedimentary systems, it is improbable that meteoric water could have added this much aluminum to ironstones. Instead, it must have been present in detrital clays.

Another hypothesis is that the ooliths themselves are detrital, being eroded soil ooliths and pisoliths from lateritic terrains (Siehl and Thein 1978, Nahon and others 1980). This is an attractive idea, because it provides a source of the iron and the aluminum and an explanation of the texture. Most deposits that I have studied, however, have nuclei for the ooliths incompatible with a soil origin, such as phosphatic or calcareous shell debris or quartz grains; furthermore, soil ooliths and pisoliths typically exhibit a radial orientation of clay flakes, whereas ironstone ooliths are tangential (Bhattacharyya and Kakimoto 1982). Another good test for the presence of ooliths from lateritic soils is the presence of high-aluminum goethite (Nahon and others 1980, p. 1295), which is generally absent from ironstones (Fig. 2-16). Thus, although it might apply in some cases, this cannot be a general explanation. These arguments do not, however, preclude a pre-concentration of the iron in the source area by lateritic weathering, with rearrangement into ooliths in the environment of deposition.

A more likely explanation for the origin of the ooliths is that of Knox (1970) who suggested that the chamositic ooliths formed in relatively quiet water from which they were washed up into shoals, possibly by storms. Most likely, the goethitic ooliths were formed in more agitated water (James and VanHouten 1979). Alternating bands of chamosite and goethite can be explained by reworking of the ooliths from the less agitated part of the depositional environment into more agitated water, then back again. As discussed in the geochemistry section, it is unlikely that the goethite ooliths arise by oxidation of pre-existing chamosite.

But where does the iron come from? There are five ways in which iron can be transported to the site of deposition (Table 2-12). As yet there is not sufficient

Table 2-12. Possible mechanisms for the production of oolitic ironstones.

Origin of Ooliths
  Replacement of pre-existing aragonite ooliths by iron-bearing solutions.
  Erosion and redeposition of soil ooliths and pisoliths from lateritic terrains.
  Direct precipitation from seawater as true ooliths.
  Formation in a quiet water environment, either as scattered ooliths on mudflats or as
    microconcretions within the sediment, then transport into sand bars.

Source of the Iron
  Volcanic
  Ground water
  Downward—from overlying organic-rich muds.
  Upward—from underlying aquifers such as submarine springs.
  River water (iron in true solution, as colloids, or as organic complexes).
  Seawater (upwelling of deep, anoxic basin water onto an aerated shelf).
  Pore waters (upwards diffusion of iron from underlying organic-rich sediments).

information to point to any one of these as the correct mechanism, but some are more likely than others. For example, there is no evidence for a volcanic source for the iron, and supply of iron by diffusion from underlying organic-rich sediments is unlikely, based on the observed vertical sequence, although non-oolitic iron accumulations with this presumed origin can be found in the Gulf of Mexico (Pequegnat and others 1972) and the Black Sea (Manheim and Chan 1974, p. 171).

The other sources have all been advocated at one time or another. Downward-moving ground water, Kimberley's (1979) model, seems unlikely for reasons already mentioned, but upward-moving ground water, discharging from submarine springs, could introduce iron to the depositional environment to be precipitated as primary ooliths, thus avoiding the difficulties with replacement of earlier carbonate ooliths. Bog iron ores form in a similar fashion, which led James (1966, p. W48) to suggest this as a general mechanism for ironstones. It has the great advantage of providing for introduction of ferrous iron into an agitated environment, without simultaneously introducing large amounts of clastics. Submarine springs are common on modern shelves (Manheim and Chan 1974, p. 168), but not associated with any noticeable iron enrichment. If ground water were important in genesis of ironstones, it should be detectable in their oxygen isotopes. For siderite, and even more so for chamosite and goethite, we do not yet know how to interpret oxygen isotopes, but they should be measured, and calcite cements could be compared with those in carbonate rocks (e.g., Magaritz and others 1979).

Transport of iron by river water has probably been the most popular hypothesis (e.g., Taylor 1949, p. 80; Hallam 1975, p. 44). Very little iron can be transported in true solution under oxidizing conditions, and Boyle and others (1977) have shown that almost all of the "dissolved" iron in rivers is carried as mixed iron oxide-organic matter colloids. These are rapidly precipitated in estuaries, and so constitute a potential source of iron for semi-restricted, near-shore basins. An alternative mode of transport of iron is as coatings on detrital clays (Carroll 1958) or as laterite particles. Such iron can be dissolved, under reducing conditions, to be reprecipitated elsewhere in the basin. Lemoalle and Dupont (1973) suggested that iron for oolith formation enters Lake Chad as clay coatings.

Another source of iron is upwelling of basinal water onto the shelf (Borchert 1960), as discussed for the iron-formations. For Phanerozoic oceans, the sulfur content has probably always been too high for this to be a general mechanism: the Black Sea today does not contain appreciable dissolved iron because it is mostly precipitated as a sulfide in deep water. During parts of the Pliocene and Miocene, however, the Black Sea was brackish to fresh water, and large siderite deposits formed (Hsu 1978a, p. 513; 1978b). Thus, Borchert's model might apply to large brackish-water basins, but most ironstones seem to have been deposited in water of normal marine salinity.

Further insight into the feasibility of these mechanisms can be gained by considering the amount of iron each is likely to be able to deliver. For a standard of comparison, we can use the Kerch ironstone (Pliocene) of the Soviet Union. From data in Kimberley (1979, p. 115 and 121), iron was deposited here at a rate of 10^8 kg/yr, close to that estimated by Holland (1973) for the Precambrian IF of the Hamersley basin. For river transport in solution, we can use the Mullica River of New Jersey as typical of the highest likely iron concentrations; it contains 25 micromoles/liter (Boyle and others 1977, table 5). Thus, a deposit like Kerch would require a river or group of rivers with a discharge of 7×10^{13} liters/yr, similar to the Rhine or the Nile. But such a high discharge is normally associated with suspended sediment loads of between 10^6 and 10^8 tons/yr (Strakhov 1967, table 3). Thus, the ratio of clastic to iron transport would be at least 10 to 1. Consequently, if the iron for these deposits is introduced as colloids by rivers, an exceptionally efficient mechanism must exist for separating the iron from the clastics. For river transport of iron in the solid phase, there is little data on which to base a calculation, but some information is available for the Nile, which we can perhaps take as representative of large tropical streams. It has a suspended solids concentration of about 540 ppm, of which about 9 percent is iron (Garrels and Mackenzie 1971). Thus, at a discharge of 3000 m³/sec, it contributes 40×10^8 kg of Fe to the Mediterranean each year. Its SiO_2 concentration is too high for this sediment to be similar to an ironstone, about 45 percent, but it can be seen that it would be feasible for a deeply weathered, low-quartz terrain to contribute large amounts of particulate iron to a marine basin.

For submarine springs of fresh water, the highest iron concentration is likely to be about 10 ppm (Strakhov 1967, table 16). Accordingly, a discharge of 7×10^{12} liters/yr would be required to form a deposit like Kerch. Data from coastal plain aquifers suggests that this volume would be hard to obtain. The Castle Hayne Limestone, the principal confined aquifer of the North Carolina coastal plain, transports water seaward into estuaries at 4 to 6×10^{10} liters/yr (calculated from data in Sherwani 1973, p. 27-28); porous shallow aquifers of the Dutch coastal plain transmit water at a rate of 25 m/yr (deVries 1974), which, for an average aquifer thickness of 100 m and a discharge width of 100 km, would result in only 10^{11} liters/yr discharged to the oceans. Thus, this mechanism seems to be adequate only for smaller deposits.

Upwelling of deep basin water can supply considerably more iron. To use the Black Sea as an example, it receives 10×10^6 tons/yr of suspended sediment, mostly from the Danube (Strakhov 1967, table 3). If only one percent of this amount were present as reactive iron compounds (data in Berner 1971, p. 208 suggest as much as 6 percent), then 8×10^8 kg/yr of Fe would be available. Thus, this mechanism is quantitatively adequate, but probably is not generally applicable because it requires brackish or fresh water.

To summarize, there is no generally acceptable mechanism for the formation of these deposits. Possibly they are polygenetic. At present, supply of iron as clastic particles by streams draining deeply weathered tropical terrains seems the most plausible. The ooliths probably formed from this iron as micro-concretions in mudflats (chamosite) or as true ooliths in more agitated water (goethite).

CHAPTER 3.

Copper and Silver

Now the fiery blood began to dissolve the sword in iron icicles. It was wonderful how it melted completely away, like ice when the Father who controls the tides and seasons looses the frozen shackles and frees the imprisoned waters.

Beowulf, line 1605-1610.
From David Wright (trans.), *Beowulf*.
(Baltimore, Maryland: Penguin Books, 1957, p.65.)
Reprinted with permission of the publisher.

Sedimentary deposits of copper are common, contributing between 25 and 30 percent of the world's production (Jacobsen 1975, table 1). Most of this amount, however, comes from one district, the Central African Copperbelt. Silver is almost always associated with the copper in these ores, and so will also be considered in this chapter. Also discussed are supergene deposits of copper and silver. Although of diminished economic importance today, they serve to illustrate the geochemical processes involved in ore-formation without some of the uncertainties associated with interpretations of the genesis of the larger deposits.

Copper has a fairly uniform distribution in the intermediate to basic igneous rocks (Table 3-1), and Cu deposits are common in association with volcanic and plutonic rocks of these compositions. Among sedimentary rocks, carbonates are conspicuously low in Cu, which is reflected in the nearly exclusive association of commercial deposits with clastic rocks. Also, note that carbonaceous shales have nearly three times the amount of Cu found in other shales, showing the importance of reducing conditions for precipitating Cu in sediments. By far the highest concentration is in pelagic clays, where the Cu is associated with Co and Ni in iron-manganese nodules. These deposits may be an important source of Cu in the future. Silver has much the same distribution as Cu, but is not as strongly depleted in seawater, carbonate rocks, and quartz arenites.

Mineralogy

The mineralogy of Cu has been extensively studied and is relatively straightforward, although some complexities are found in sedimentary deposits because of the presence of metastable phases. For the most part, only sulfides and native metals are important economically, although carbonates such as malachite and

Table 3-1. Abundance of Cu and Ag in common geological materials.

	Cu (ppm)	Ag (ppb)	Cu/Ag × 1000
Igneous Rocks			
peridotites	47	60	0.78
basalts	90	100	0.90
andesites	53	80	0.66
granites	13	37	0.35
Sedimentary Rocks			
shales			
average	35		
red		150	
green		190	
black	95	290	0.33
sandstones			
quartz arenites	30		
arkoses			
graywackes	11	250	0.044
limestones	6	125	0.048
Sediments, seawater			
pelagic clays	251		
seawater	0.0015	0.32	0.0047

Source: Wedepohl 1969, tables 29-E-6, 29-I-2, 29-K-2, 47-E-1, 47-I-2, 47-K-1.

azurite are common and even make up the bulk of a few deposits. The usual minerals encountered are listed in Table 3-2, along with the most helpful properties for distinguishing them in polished section.

The Cu-S system was once thought to comprise only CuS and Cu_2S, but a number of intermediate phases are now known. Their mineralogy and thermochemistry has been researched in detail, but less is known about their distribution in nature. Of the chalcocite-like phases, anilite and djurleite are optically indistinguishable from chalcocite; careful X-ray or microprobe study is necessary to identify them. Digenite can usually be distinguished from chalcocite, where the two occur together, by its more blueish color. Digenite apparently requires small amounts of iron in its structure (Potter 1977, p. 1531), and so may be an intermediate phase in the transformation of chalcocite to bornite or vice versa.

Covellite also occurs in several forms. Those which fail to show the characteristic blue to red bireflectance in oil are termed blaubleibender (blue-remaining) covellites. At least two distinct compositions occur (Table 3-2), for which the mineral names yarrowite and spionkopite have been used (Goble 1981, table 1). These phases appear to form as metastable intermediates in the reduction of normal covellite to chalcocite (Rickard 1972):

$$2CuS + 2H^+ + 2e^- \rightarrow Cu_2S + H_2S,$$

or the reverse, the oxidation of chalcocite to covellite during supergene enrichment (Sillitoe and Clark 1969, p. 1703).

Table 3-2. Common sulfides and native metals of copper-silver deposits.

Name	Formula	Color (in oil)	Bireflectance	Anisotropy
Chalcocite	Cu_2S	Bluish white	Very weak	Weak: green to light pinkish
Djurleite	$Cu_{1.934}S - Cu_{1.965}S$		Same as chalcocite	
Anilite	$Cu_{1.75}S$		Same as chalcocite	
Blaubleibender covellite	$Cu_{1.1}S$ or $Cu_{1.4}S$	Indigo blue	Extremely high: Air: O–blue E–bluish white Oil: O–deep blue E–bluish white	Strong: orange to red
Covellite	CuS	Indigo Blue	Extraordinarily high: Air: O–deep blue E–bluish white Oil: O–violet red E–blue gray	Very strong: fiery orange
Digenite	$Cu_{1.8-x}Fe_xS$	Grayish blue	Not present	Isotropic
Bornite	Cu_5FeS_4	Purple	Very slight	Very weak
Idaite	Cu_3FeS_4	Red	Very strong: O–reddish orange E–bright yellow gray	Extremely strong: vivid green in 45° position
Chalcopyrite	$CuFeS_2$	Brassy yellow	Weak	Weak but distinct: gray-blue to greenish yellow
Pyrite	FeS_2	Yellowish white	Not present	Isotropic
Marcasite	FeS_2	Yellow	Strong: white to yellow	Strong: blue, green-yellow or purple
Native copper	Cu	Brownish pink	Not present	Isotropic
Native silver	Ag	Bright white to pinkish	Not present	Isotropic
Acanthite	Ag_2S	Gray	Very weak	Weak, but distinct

Chemical compositions are from Potter (1977); optical properties from Uytenbogaardt and Burke (1971).

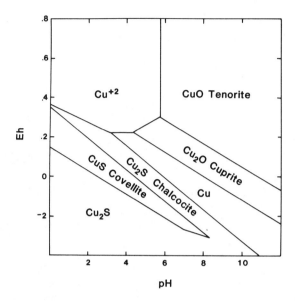

Fig. 3-1.
Eh-pH diagram for copper oxides and sulfides showing region of solubility under oxidizing conditions at pH less than 6 (ΣCu, $\Sigma S = 10^{-4}$).

Geochemistry

Eh-pH changes and chloride complexing dominate the geochemical behavior of Cu and Ag at low temperatures. As shown in Fig. 3-1, there is a large field of Cu solubility under oxidizing conditions at moderate to low pH. Iron is less soluble under these conditions (Fig. 2-4), so that an efficient mechanism is available for separating Cu from Fe. Also, note that Cu should precipitate as a sulfide or as native copper under reducing conditions. Thus, Cu will tend to

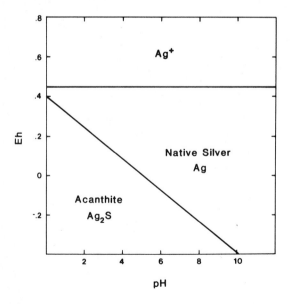

Fig. 3-2.
Eh-pH diagram for Ag shows, as with Cu, a region of solubility at high Eh, but no insoluble oxides at high pH ($\Sigma Ag = 10^{-6}$, $\Sigma S = 10^{-2}$).

Fig. 3-3.
Effect of Cl$^-$ on Cu solubility.
Note the region of increased
solubility at Cl$^-$ > 10^{-2} (pH =
5, ΣCu = 10^{-4}, ΣS = 10^{-4}).

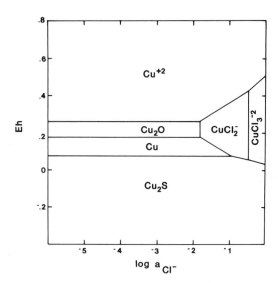

migrate from areas of oxidation and concentrate in areas with reducing conditions.
The redox behavior of Ag is similar (Fig. 3-2) except that there are no Ag oxides
or carbonates analogous to those of Cu. Because of this similarity, Ag tends to
parallel Cu in enrichment. Most Cu deposits also produce Ag, and Cu/Ag ratios
in ores are close to that for the Earth's crust (Rose 1976). An important exception
is the Central African Copperbelt, which has Co rather than Ag as an accessory.

Chloride-bearing solutions are important, perhaps critical, for the transport of
these metals because their solubility is greatly increased by chloride complexing
(D.E. White 1968). Rose (1976) has shown that the solubility of Cu in particular
is greatly enhanced by the formation of such complexes as $CuCl_2^-$ and $CuCl_3^{2-}$
(Fig. 3-3, 3-4). In fact, transport of sufficient Cu to form large orebodies in the
absence of chloride complexing is difficult, if not impossible, unless some pre-
existing enrichment, such as in supergene alteration of porphyry coppers, is
present.

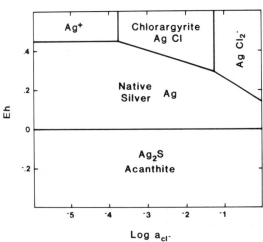

Fig. 3-4.
Cl$^-$ has a more complex effect
on Ag, first causing chlorar-
gyrite to form, then at higher
concentrations, favoring dissolution of Ag because of the formation of the bichloride complex
(pH = 5, ΣAg = 10^{-6}, ΣS =
10^{-2}).

Supergene Enrichment

The geochemical properties outlined above can be illustrated by considering the behavior of Cu and Ag during supergene enrichment. Most Cu deposits, particularly hydrothermal deposits of the porphyry type, exhibit some degree of enrichment near the water table. Formerly, these enrichment zones were the principal source of Cu, but now the supergene Cu in most deposits is exhausted, and the lower-grade primary material is mined. Note that these ores are normally discussed along with their hydrothermal protores, but the enrichment itself is entirely a surficial process. Furthermore, it serves as a useful comparison for Cu deposits in sediments, and as a useful contrast to other metal enrichments in soils such as Al and Ni (Chapter 4).

These enriched zones display a pronounced vertical zonation. There are two major subdivisions: the overlying zone of oxidation, and, below the former position of the water table, the zone of enrichment. The zone of oxidation is characterized by large amounts of goethite, accompanied in some cases, especially where a carbonate wall rock is present, by oxidized Cu minerals such as malachite or tenorite. Under very arid conditions, as in the Atacama desert of Chile, chlorides (atacamite) or sulfates (brochantite) are also found in the zone of oxidation. The Chilean deposits have produced large amounts of Ag, much of it from halides in the zone of oxidation, but most from native Ag and Ag sulfides in the zone of enrichment (Park and MacDiarmid 1975, p. 487-491). In Cu enrichment, the chief ore mineral reported is "chalcocite" (see, for example, the descriptions of individual deposits in Titley and Hicks 1966), much of which is probably djurleite and anilite. Some covellite is also found, usually overlying the chalcocite at the boundary between oxidizing and reducing conditions (Peterson 1954, p. 374).

The broad division into an overlying leached, oxidized zone and an underlying, enriched reducing zone is readily explained by Fig. 3-5, in which the Eh-pH diagram for iron has been superimposed on that for Cu. Rainwater, falling on a pyrite-containing rock or a soil derived from it, will begin with a composition like that given by point A, but will move rapidly to a lower pH owing to the oxidation of the ferrous sulfide. As this water passes down through the soil, it leaches Cu from primary Cu minerals such as chalcopyrite or from earlier supergene Cu minerals, leaving behind the Fe as goethite. When this water reaches the ground water table, it encounters reducing conditions, and precipitates its Cu, to some extent as native copper, but mostly as "chalcocite." Note that the Fe is now soluble, so there will be a tendency for "chalcocite" to replace pyrite, but because pyrite is relatively slow to react, bornite and chalcopyrite are the minerals preferentially replaced (Anderson 1955, p. 327). It can be seen from Figs. 3-2 and 3-4 that Ag should be subject to the same behavior.

Similar diagrams can be used to show why this process is only important for Cu and Ag (Garrels 1954). Lead has an insoluble sulfate, anglesite, and so is left behind in the zone of oxidation in a dispersed state. Zinc, on the other hand, is relatively soluble under all conditions, and so tends to be lost from the system

Fig. 3-5.
During weathering there will be a tendency for Fe to be immobilized at the surface as Fe hydroxides, forming a gossan, while the more soluble Cu is leached downwards to be precipitated as chalcocite or perhaps as native Cu at the water table.

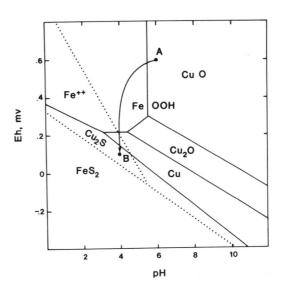

altogether. An exception can be found in the limestone-hosted lead-zinc deposits of Mexico, where the leached zinc is redeposited lower in the profile as the carbonate, smithsonite (Jensen and Bateman 1979, fig. 14-3).

This simple Eh-pH model fails, unfortunately, to explain the details of the sulfide mineralogy. For example, the various metastable phases are unaccounted for. Furthermore, covellite should be found under more strongly reducing conditions than chalcocite, based on their positions on Fig. 3-1, yet it is normally found at the boundary between the enriched and oxidized zones. Experimental evidence (Sato 1960) indicates that oxidation of chalcocite does not proceed by the simple congruent reaction used to calculate the Eh-pH relations shown,

$$Cu_2S + 4H_2O \rightarrow 2Cu^{2+} + SO_4^{2-} + 8H^+ + 10e^-.$$

Instead, there is a step-wise removal of Cu that leaves a solid progressively enriched in sulfur. For instance,

$$Cu_2S \rightarrow Cu_{1.75}S + 0.25Cu^{2+} + 0.50e^-$$
$$Cu_{1.75}S \rightarrow CuS + 0.75Cu^{2+} + 1.50e^-$$
$$CuS \rightarrow Cu^{2+} + S + 2e^-.$$

Note that in this case djurleite and covellite form as metastable intermediates in the oxidation of chalcocite, but, because they are not the phases with the lowest free energy under these conditions, they do not appear on thermodynamic diagrams. There is some field evidence for this sequence. Sillitoe and Clark (1969) have described the oxidation of a supergene Cu deposit in Chile in which progressions like

$$\begin{array}{ccccc} Cu_{1.96}S & \rightarrow & Cu_{1.4}S & \rightarrow & CuS \\ \text{(djurleite)} & & \text{(blaub.cov.)} & & \text{(covellite)} \end{array}$$

are found. During reduction, within the zone of enrichment, the reverse should happen. Cailteux (1974, p. 272-275) reported a step-wise addition of Cu during supergene enrichment of sedimentary Cu in Zaire,

$$CuFeS_2 \rightarrow CuS \rightarrow Cu_{1.4}S \rightarrow Cu_{1.8}S \rightarrow Cu_{1.95}S.$$

If these sequences prove to be generally applicable, there should be two generations of covellite in a supergene deposit, one at the top of the enriched zone replacing chalcocite, the other at the base of the enriched zone replacing chalcopyrite or bornite.

Examples

Let us now turn to an examination of four well-known occurrences of Cu in sedimentary rocks to see how the principles just developed can be applied. Of the various types (Table 3-3), sandstone Cu is similar to roll-front uranium mineralization (Anhaeusser and Button 1972, Shockey and others 1974, Woodward and others 1974, Caia 1976), and so the discussion in Chapter 6 covers the principles of their genesis. The others are treated in the order of increasing controversy in their interpretation.

White Pine, Michigan

The upper peninsula of Michigan contains two sizeable Cu districts (Brown 1974). The first contains native copper and is found in the basic volcanics of the Portage Lake lava series; the second, in the base of the Nonesuch Shale, is mostly chalcocite with subordinate native copper. Several hundred meters of red beds, the Copper Harbor Conglomerate, separate the two stratigraphically. The native copper deposit may be magmatic (Amstutz 1977) or hydrothermal-epigenetic (W.S. White 1968), but the ores in the Nonesuch Shale are most

Table 3-3. Varieties of sedimentary copper deposits.

Type	Example	Dominant Process
Supergene	Chincharilla, Chile	Weathering of pre-existing accumulation.
Epigenetic in sandstone	Paoli, Oklahoma	Introduction by lateral ground water flow.
Epigenetic in shale	White Pine, Michigan	Introduction by upwards ground water flow.
Red-bed evaporite	Creta, Oklahoma	Introduction by saline ground water in sabkhas.
Controversial	Kupferschiefer, Central African Copperbelt	Variously interpreted as syngenetic, or one of the above, or even hydrothermal.

likely ground water-epigenetic. A low-temperature origin of these sulfides is suggested by the predominance of chalcocite and djurleite and by preservation of temperature-sensitive organic constituents (Brown 1971, p. 565); epigenetic rather than syngenetic deposition of the Cu is indicated by the discordance of the mineralization with bedding on a regional scale, and by the mineral zonation (White and Wright 1966).

The zoning has been studied in detail in the area near the town of White Pine by Brown (1971). The sequence, passing upward, shows a steady decrease in Cu and an increase in iron:

native copper → chalcocite → bornite → chalcopyrite → pyrite

(Fig. 3-6). The pyrite zone makes up the bulk of the Nonesuch, and contains no more that 0.05 percent Cu. Other minerals found are djurleite and digenite at the top of the chalcocite zone, suggesting a gradual transition from chalcocite to bornite. Sometimes, native silver occurs with the native copper; covellite is conspicuously rare. Note that the sequence of minerals found is nearly the same as that seen in supergene enrichment, but upside down, indicating that Cu was introduced from below.

A probable source of this Cu is Cl-rich ground water flowing through the underlying Copper Harbor Conglomerate. This unit thins sharply under the area of mineralization (Fig. 3-7), a geometry which is thought to have caused a portion of the ground water discharge in the conglomerate to pass upward through the Nonesuch (W.S.White 1971). The Nonesuch is actually a siltstone rather than a shale (Ensign and others 1968, p. 475), and so would have had a higher permeability than the formation name implies. Copper in the water could have reacted with the early diagenetic pyrite in the Nonesuch to produce the sequence of minerals found. The ultimate source of the Cu is not known, but it is reasonable

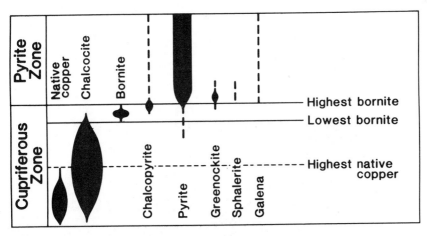

Fig. 3-6. Mineral zoning at White Pine is similar to that found in supergene enrichment, but inverted, which suggests that Cu-rich water passed upwards (after White and Wright 1966, fig. 3).

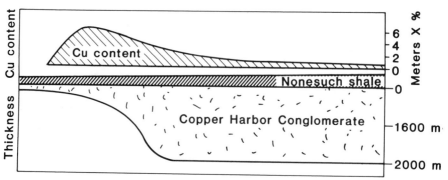

Fig. 3-7. Cu content in the Nonesuch Shale is inversely related to the thickness of the underlying Copper Harbor conglomerate, the probable conduit for the Cu-rich fluids. Horizontal axis represents distance to the north from the outcrop (after W.S. White 1971, fig. 2).

to suppose that it is in some way related to the native copper in the underlying volcanics. Perhaps oxidizing ground waters leached Cu from the volcanics; perhaps chlorides participated in the mobilization (D.E.White 1968, p. 324); it is even possible that hydrothermal fluids discharged into the Copper Harbor, then moved updip to deposit their Cu in the Nonesuch (Brown 1981).

Sulfur isotopes show a wide range of values (Fig. 3-8), consistent with a syngenetic or early diagenetic origin of the sulfur. There is extreme variability, both within and between beds: samples with $\delta^{34}S$ of -10 and $+20$ permil can be separated horizontally by only a hundred meters. Burnie and others (1972) suggested that this variability was a result of deposition of the Nonesuch in a nearshore, perhaps brackish, environment in which the rate of supply of sulfate was variable. Unfortunately, the only data are from the ore zone; the bulk of the shale host has not been examined, so that the possibility of fractionation during replacement of the iron sulfides by Cu, or of introduction of small amounts

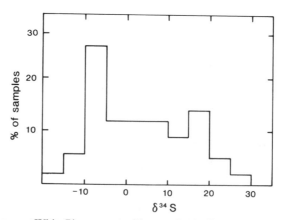

Fig. 3-8. S isotopes at White Pine cover a wide spread, possibly owing to local differences in the amount of sulfate from seawater (after Burnie and others 1972, fig. 4d).

of sulfur of a different isotopic composition by the Cu-bearing fluids, cannot be excluded.

Creta, Oklahoma

Chlorides are intimately associated with Permian copper deposits of Oklahoma, and are thought to have been important for metal transport. The mineralization is found in a near-shore, red bed-evaporite sequence (Fig. 3-9), with the highest Cu concentrations occurring in gray shales immediately overlain by beds of gypsum (Hagni and Gann 1976, p. 40). Most of the Cu-bearing shales were deposited on tidal flats (Smith 1976).

The ores consist of digenite with subordinate chalcocite (including djurleite and anilite?). Chalcopyrite and native copper are occasionally seen, and covellite is found replacing chalcocite (Kidwell and Bower 1976, p. 52-53). Microprobe study of subeconomic Cu mineralization in equivalent strata in Kansas has re-

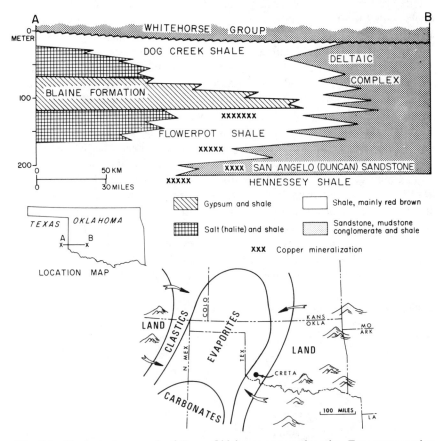

Fig. 3-9. Geologic framework of Texas-Oklahoma copper deposits. *Top:* cross-section showing position of deposits in fine clastics between evaporites and coarser deltaics. *Bottom:* inferred paleogeography (Johnson 1976, fig 3; Dingess 1976, fig. 2).

vealed the presence of anilite- and djurleite-like phases, but no chalcocite (Table 3-4). Some digenite was also found as intergrowths in bornite. The chalcocite-like phases have two compositions, one close to anilite ($Cu_{1.75}$) and one intermediate between anilite and djurleite ($Cu_{1.934}$). The second is thought to be due to fine intergrowths of anilite in djurleite (Ripley and others 1980, p. 727).

Most workers agree that the sulfur was deposited as syngenetic or early diagenetic iron sulfides, often replacing organic matter, and that these were subsequently replaced by Cu sulfides. Sometimes, the Cu minerals directly replaced organic matter, particularly the spore *Triletes* (Hagni and Gann 1976) and wood fragments (Smith 1976, p. 33).

Sulfur isotopes, as at White Pine, show a wide spread, from -7 to $+35$ permil (Lockwood 1976, p. 33). There are not enough data, however, to say whether this variation has any relationship to depositional environments.

Smith (1976) has related the introduction of the Cu to the presence of the evaporites, using a model proposed by Renfro (1974). Noting that sedimentary Cu deposits are commonly arranged in the vertical sequence red beds → dark shale with Cu → evaporites, Renfro proposed that Cu was introduced after deposition of the shale by the peculiar ground water circulation of sabkha environments (Fig. 3-10). The high rate of evaporation in coastal desert areas leads to a strong upward flow of ground water in the tidal zone, with subsequent precipitation of gypsum and halite. Renfro theorized that, if the right geometry of facies were present, this ground water could leach Cu from underlying sands, in the process forming red beds by oxidizing the iron, then precipitate the Cu as sulfides in the algal mat layer that commonly underlies sabkha evaporites (Fig. 3-11). Thus, the sulfur would be syngenetic, or early diagenetic, but the metals would be added during later diagenesis, bordering on ground water-epigenetic. Both Smith (1976, p. 35) and Renfro (1974, p. 41) believed that the water carrying the Cu came from the terrestrial side of the sabkha, because of the high Eh and low pH of meteoric water compared with seawater. Rose (1976, p. 1046), on the other hand, has argued that Cl concentrations would still be too low, and contribution of chlorides from the marine side of the ground water system is required. Ideally, then, what is needed is chloride-bearing water, of high Eh and low pH, derived from the landward side. Such a situation has been described

Table 3-4. Average microprobe analyses of copper sulfides from Permian red-beds of Kansas.

	Cu	Fe	S
Chalcopyrite	1.00	0.99	2
Bornite	4.62	0.95	4
Digenite	1.65	0.07	1
"Chalcocite"			
Anilite	1.78	—	1
Djurleite	1.91	—	1

Source: Ripley and others 1980, p. 726–727.

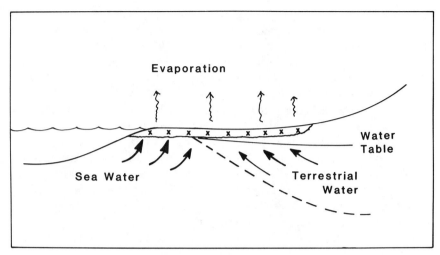

Fig. 3-10. Circulation in the sabkha environment. Evaporation induces an upwards flow of water, that can be supplied from the sea or the land side, from which evaporite minerals are precipitated.

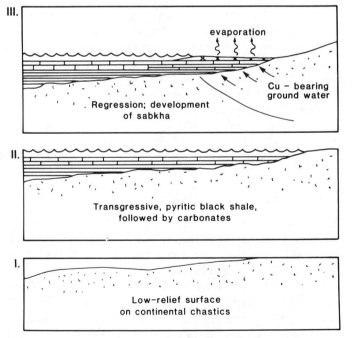

Fig. 3-11. Model of Cu mineralization in a sabkha environment. Sulfur is deposited during early diagenesis of the black shale unit of stage II, whereas the Cu is deposited epigenetically by ground water passing upwards through the shale driven by the sabkha circulation (after Renfro 1974, fig. 6).

from a sabkha in Libya (Rouse and Sherif 1980), where sulfur isotopes suggest that terrestrial ground water greatly predominates over marine, but is relatively saline because of dissolution of older evaporites. The involvement of evaporites in the deposition of Cu had earlier been proposed by Davidson (1966), but he envisioned them as a source of Cl for deeply circulating basinal brines, rather than as an intimate part of the depositional system, as in Renfro's model.

Note that the circulation pattern illustrated in Fig. 3-11 should produce a mineral zoning like that at White Pine, with the minerals richest in Cu at the base. Little if any zoning is developed in the rocks at Creta, however: the sulfides are confined almost entirely to intergrowths of digenite and "chalcocite." Kidwell and Bower (1976, p. 53) reported some chalcopyrite from shales a few feet above the main "chalcocite" mineralization, which would be in the proper sequence, and Ripley and others (1980, p. 725) reported a sequence from Kansas with "chalcocite" overlain by bornite-chalcopyrite, the expected order, but pyrite is common just below the "chalcocite," inconsistent with an introduction of Cu from below.

The Kupferschiefer

Although evidence is strong for post-depositional addition of Cu to the Permian shales of the U.S., the source of metals in the European deposits of the same age has been more controversial. The Kupferschiefer contains what is probably the world's most famous sedimentary ore deposit. This extremely thin—less than one meter—black, dolomitic shale covers most of northern Europe, and has been mined for Cu, Pb, and Zn for centuries. The stratigraphic sequence is similar to that in the North American deposits (Fig. 3-12), suggesting a comparable origin. Many workers, however, have regarded the Kupferschiefer as the best example of a syngenetic ore deposit (e.g., Wedepohl 1971).

Unlike the Permian deposits of the U.S., the Kupferschiefer is rich in Pb and Zn as well as Cu (Fig. 3-13). Further, it has a pronounced mineral zoning, both vertically and horizontally. Passing upward, the Cu-Fe minerals are distributed in the sequence

hematite → chalcocite-digenite-covellite → bornite → chalcopyrite → pyrite

much like that seen at White Pine. In this case, commercial quantities of Pb and Zn sulfides appear in the bornite zone, and the zonation is repeated horizontally, passing from east to west (Jung and Knitzschke 1976, p. 375).

Wedepohl (1964, 1971), arguing against a hydrothermal-epigenetic source of the metals, constructed a syngenetic model for the deposits. He pointed out that the large-scale mineral zonation, plus the confinement of the mineralization to a narrow stratigraphic interval, are inconsistent with a post-depositional introduction of the metals. He postulated that the Cu, Pb, and Zn, as well as the trace elements in which the shale is also enriched (V, Cr, Mo), were leached from the underlying Rotliegendes sands, then carried by streams into the basin of deposition where they were deposited syngenetically. A similar mechanism

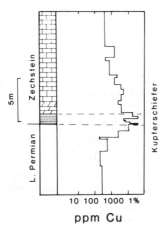

ppm Cu

Fig. 3-12. Stratigraphic section of Permian rocks in Poland. Overlying the limestones (Zechsteinkalk) is a thick evaporite sequence (after Haranczyk 1970, fig. 6).

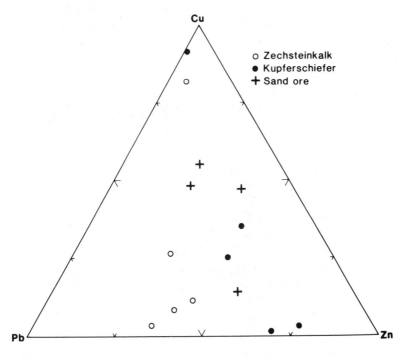

Fig. 3-13. Unlike other sedimentary Cu deposits, the Kupferschiefer is also enriched in Pb and Zn. Note the predominance of Zn in the Kupferschiefer itself, Pb in the overlying Zechsteinkalk. (Data from Rentzsch 1974, table 4.)

has been advocated by Oberc and Serkies (1968), with the addition of a stage of diagenetic remobilization of the metals. Dunham (1964) proposed a syngenetic deposition of metals supplied by hot springs.

A host of models involving post-depositional processes have been proposed (Tourtelot and Vine 1976, p. C16), and detailed work on the eastern deposits indicates that a diagenetic mechanism is more likely than a strictly syngenetic one. For example, Rentzsch (1974) has shown that, viewed on a large enough scale, the mineralization is, in fact, discordant to bedding (Fig. 3-14), and, further, that it surrounds areas of post-depositional alteration of the underlying sands, referred to as Rote Fäule (Fig. 3-15). The Rote Fäule are areas of secondary reddening that extend in places up into the evaporites, and cut across the bedding at an angle of a degree or so. All of the economic mineralization in the German Democratic Republic is associated with this facies, although there are extensive areas of Rote Fäule to the north with no mineralization. Probably, the metals were carried by ground water, possibly chloride-rich, that passed through these areas and then deposited the metals on contact with the sulfide-rich muds that now make up the Kupferschiefer. A similar mechanism has been suggested for deposits of this age in the Soviet Union (Lurje 1977).

Sulfur isotopes once again show a large spread, but in this case have a definite vertical trend (Fig. 3-16). By contrast, there do not seem to be any significant lateral variations. The vertical profile is best interpreted as reflecting an increasing isolation of the basin of deposition from contact with the ocean (Marowsky 1969, Maynard 1980). If the renewal of salts in the basin is restricted, heavy sulfur (^{34}S-enriched) will accumulate because of preferential removal of ^{32}S by bacteria

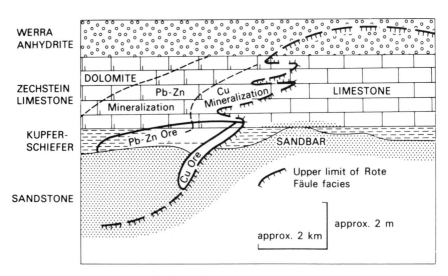

Fig. 3-14. Mineralization in the Kupferschiefer is discordant to bedding. Note also the pinching out of the shale over the high in the underlying sandstone (Evans 1980, fig. 13.2; after Rentzsch 1974, Fig. 10).

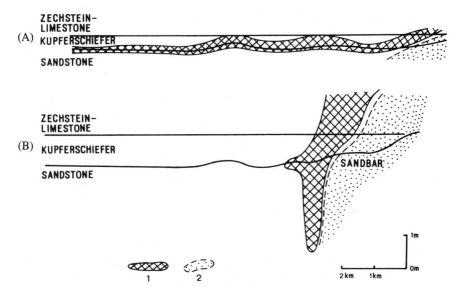

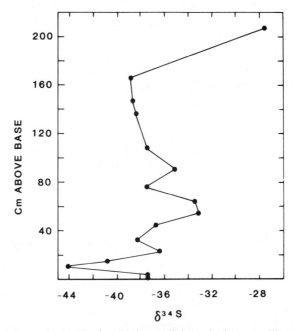

Fig. 3-15. Relationship of Cu mineralization (1) to areas of Rote Fäule (2), in Mansfeld-Sangerhausen (a) and Lusatia-Richelsdorf (b) (Rentzsch 1974, fig. 11).

Fig. 3-16. S isotopes in the Kupferschiefer are light at the base, oscillate through intermediate values, then become relatively heavy at the top. This trend may be due to increasingly closed communication with seawater, preliminary to the onset of evaporite deposition (after Marowsky 1969, fig. 3a).

during sulfate reduction. Then, because their sulfate supply is becoming increasingly heavy, the bacteria produce progressively heavier sulfides.

Central African Copperbelt

One of the world's largest Cu districts, and perhaps the most poorly understood, is the Proterozoic Central African Copperbelt of Zaire and Zambia (Fig. 3-17). Again, there is extensive mineralization of particular stratigraphic horizons, some of which show well-developed mineral zonation. The deposits were at first regarded as hydrothermal in origin (e.g., Gray 1932), but detailed stratigraphic work has led to the view that they are sedimentary, perhaps even syngenetic (Garlick 1961, Garlick and Fleischer 1972).

The Copperbelt is separated politically and geologically into two sections (Bartholomé 1974, p. 204). In Zaire, the host rocks are unmetamorphosed and relatively flat-lying, although cut by numerous thrust faults. To the south and west, in Zambia, they are metamorphosed, from chlorite up to garnet grade, and are often highly deformed. In addition, the Zambian section was deposited over a much rougher paleotopography: the sediments unconformably overlie prominent granite hills, which were once believed to be intrusives, some of which are capped by stromatolitic bioherms. Lithologically, the Zaire section is characterized by dolomite, while the Zambian is more clastic. Within Zambia, a distinction is made between the "ore shale trend" to the southwest, where mineralization is mostly in fine-grained clastics, and the area around Mufulira, where mineralization occurs at three levels in carbonaceous sandstones interbedded with barren dolomites (van Eden 1974). Both Zaire and Zambia contain evidence of hypersalinity during deposition, magnesite being common in the Zaire dolomites, anhydrite in Zambia. The deposits have a similar mineralogy: Cu and cobalt sulfides are abundant, while lead and zinc are conspicuously rare.

Some similarities can be seen in the vertical sequence of lithologies (Fig. 3-18). A basal coarse clastic unit, perhaps of fluvial origin, is overlain by finer clastics and carbonates that contain most of the mineralization, and then by a thick, barren carbonate (Zaire) or clastic unit (Zambia). Looked at in detail,

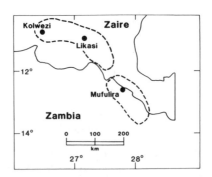

Fig. 3-17. Ores in the Copperbelt are found in two areas, one in Zaire, the other mostly in Zambia (adapted from Francois 1974, fig. 1).

however, it is obvious that, although all the sediments were deposited in near-shore environments, the depositional system was complex, and varied in character with position in the basin.

Mineralization is found at several stratigraphic levels and in a variety of lithologies across the Copperbelt. In Zaire, two zones are preferentially miner-alized (Fig. 3-18), both in dolomite or dolomitic shale. In Zambia, mineralization is found at many levels, from the footwall sandstones up to the dolomites of the hangingwall, but is best developed in the ore shale. This unit is a black to gray silty argillite, often calcareous at the base (Binda and Mulgrew 1974, p. 226-228). Mineralization occurs throughout and may be discordant to stratigraphic subunits within the shale. Stromatolitic bioherms associated with the shale are always barren, however. At Mufulira, there are three mineralized zones, which follow stratigraphic boundaries at their top, but are discordant at the base (Fig. 3-19). These zones have been related to transgressions of the sea from the southwest (Fig. 3-20).

In Zambia, the sulfides commonly display zoning, particularly around paleo-hills and bioherms (Garlick 1964, fig. 1). A typical sequence, moving outwards is

$$\text{chalcocite} \rightarrow \text{bornite} \rightarrow \text{chalcopyrite} \rightarrow \text{pyrite},$$

which Garlick interpreted as caused by an increase in the depth of water in which the sulfides were precipitated. The reverse zonation is is also common, however,

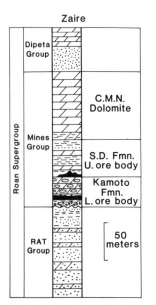
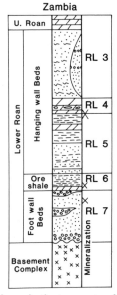

Fig. 3-18. Vertical sections through the ore horizons in the two parts of the Copperbelt. The ores in Zaire have more of a carbonate association, those of Zambia more clastic. (Data of Bartholomé and others 1973, Cahen 1974, Binda and Mulgrew 1974.)

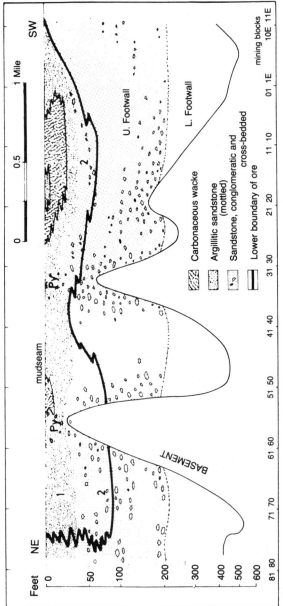

Fig. 3-19. Longitudinal section through Footwall and C orebodies at Mufulira showing that the ore, although locally concordant with bedding, is discordant when considered on a large scale (Van Eden 1974, fig. 16).

as shown in a section from Mufulira (Fig. 3-21) that is reminiscent of the distribution of minerals at White Pine. Also, Annels (1974, p. 252) has disputed the interpretation of increasing water depth on lithologic grounds. Intracrystalline zoning is also found, most prominently as an enrichment in cobalt in the rims of pyrites in the Zaire section (Bartholomé 1974, fig. 1). The significance of this zoning, or indeed of the presence of such large amounts of Co, is not known.

Stable isotopes could provide valuable information about ore-formation in this area, but the only comprehensive study that I am aware of is that of Dechow and Jensen (1965) on sulfur. The most striking feature of their data is the predominance of samples with positive delta values, especially in Zambia (Fig. 3-22). The simplest explanation for this distribution is that evaporite sulfur was an important contributor to these deposits, consistent with the presence of abundant gypsum. Note also the significantly lighter S values in Zaire, paralleling the diminished importance of evaporites in the section. Some zoning of isotope values is also found. For instance, the Kamoto deposit in Zaire becomes steadily heavier upward, from about -5 to $+5$ permil (Dechow and Jensen 1965, fig. 4). Other deposits have a reverse trend, or none, or show a lateral zonation (Fig. 3-21). At Mufulira, the pyrite is typically light, around -0.2 permil, while the Cu sulfides are uniformly heavier, averaging $+4.5$ permil. Thus, the Cu ores were not deposited by replacement of earlier, syngenetic pyrite, as envisioned for White Pine. Either some heavy sulfur was introduced along with the Cu, or anhydrite associated with the sandstone host rock was somehow reduced to sulfide during the introduction of the Cu. A thorough study of carbon and oxygen isotopes is needed before much more progress can be made on these problems.

Theories of genesis of the Copperbelt deposits fall into three categories: syngenetic, diagenetic, and epigenetic. The syngenetic view, that is, that Cu and Co were precipitated directly as sulfides at the time of sedimentation, has been developed by Garlick and co-workers (Garlick 1961, Garlick and Fleischer 1972, Fleischer and others 1976, Garlick 1981). They have been especially impressed with the relationship of mineralization to paleotopography, and its restriction to particular stratigraphic horizons. The diagenetic view, based on analogy with the Cu deposits discussed earlier in this chapter, has been advocated by Bartholomé (1974) and van Eden (1974). Bartholomé cited petrographic evidence, van Eden the local discordance of bedding and mineralization. van Eden also suggested that there was a primary Cu mineralization, possibly as much as 0.5 percent, which was enriched to present grades (3 to 6 percent) during diagenesis. Annels (1974, 1979) proposed an epigenetic model. He hypothesized that deep formation brines carried Cu upward along permeable zones, perhaps fault-related, to be trapped beneath impermeable seals, such as the Mudseam at Mufulira. Based on the antipathetic relationship between Cu and anhydrite, he suggested that the sulfur in the Cu sulfides was derived from reduction of the sulfates.

At present, we do not have sufficient information to choose among these hypotheses, although indications are that some form of post-depositional introduction of the metals is most likely. Further, the diversity of the occurrences

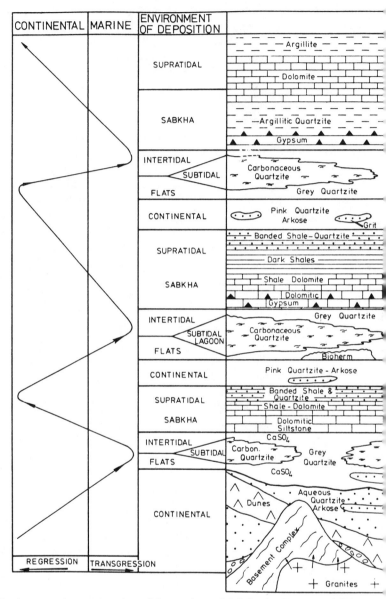

Fig. 3-20. Environmental reconstruction of the section at Mufulira by Bowen and Gunatilaka (1977, fig. 6.8).

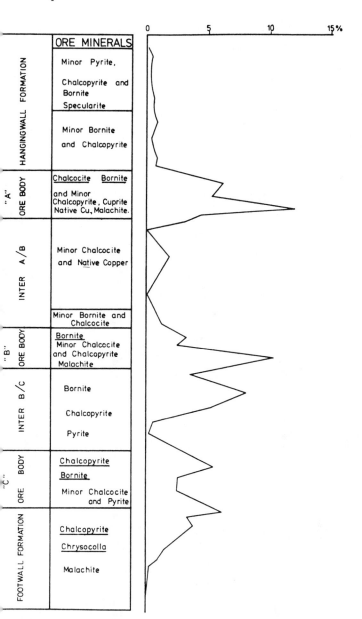

Fig. 3-20. (Continued)

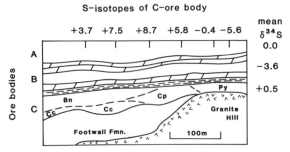

Fig. 3-21. Zoning of copper minerals and S isotopes at Mufulira, showing heavier S in
Cu minerals than in pyrite, and increasing δ values in the C-ore body away
from the paleohigh (after Dechow and Jensen 1965, figs. 3 & 11).

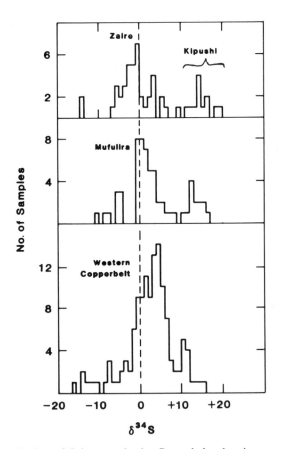

Fig. 3-22. Distribution of S isotopes in the Copperbelt, showing successively lighter
values to the NW (after Dechow and Jensen 1965, fig. 5).

suggests that more than one mechanism may have been operative. More careful stratigraphic work, directed towards finding general, as opposed to single-deposit, connections between the sedimentary sequence and mineralization are clearly called for. In particular, comparisons across the political and language barriers separating Zaire and Zambia are needed. A final problem concerning these deposits that needs more attention is the abundance of cobalt. Why is this metal so common, while Pb, Zn, and Ag are so rare?

Summary ,

Sedimentary Cu-Ag deposits are found almost exclusively in fine-grained rocks, usually dark siltstones and shales. In most cases, red beds and evaporites also make up part of the sequence. The behavior of Cu and Ag is controlled by oxidation and reduction: the metals are leached from areas of oxidation and precipitated in areas with reducing conditions. The presence of Cl^- greatly enhances solubility, and hence the transport, of the metals.

Sulfides are by far the most important minerals. Based on analogy with supergene enrichment, their zoning can be interpreted as forming the sequence

$$\text{chalcocite} \rightarrow \text{bornite} \rightarrow \text{chalcopyrite} \rightarrow \text{pyrite}$$

in the direction of transport of the Cu. By using electron microprobe analysis, other minerals in the Cu-S system can be easily detected, so that refinements in this zoning sequence are to be expected and should be sought.

Isotope studies are still in a reconnaissance stage. Only sulfur has been measured, and most workers have been content to document a high variability. Our understanding of isotope fractionation during sedimentary processes has advanced sufficiently that we should be able to explain the details of this variation, and use them to help explain the genesis of the deposits.

Ideas about genesis have shown an evolution. Most early workers regarded these deposits, along with almost all sulfide ores, as products of igneous activity. Then, after the Second World War, syngenetic models became popular. In recent years, based largely on improvements in our understanding of sedimentary processes, there has been a tendency for models involving diagenetic mineralization to replace strict syngenetic ones.

CHAPTER 4.

Aluminum and Nickel

Meanwhile the rain fell with cruel persistence. You felt that the
heavens must at last be empty of water, but still it poured down,
straight and heavy, with a maddening iteration, on the iron roof.

Rain, Somerset Maugham
From *Complete Short Stories of Somerset Maugham*
(© 1921 by Smart Set, Inc.)
Reprinted by permission of Doubleday & Co., Inc.

In the last chapter we discussed concentration of Cu in soil profiles by supergene
enrichment. Two other elements that are enriched to ore grade by soil processes
are aluminum and nickel. Virtually all of the world's Al and perhaps one third
of its Ni comes from such deposits (Lelong and others 1976, p. 147). Aluminum
ores overlie a variety of rock types, but Ni is associated exclusively with ultramafics, reflecting the relative abundance of the two elements in the crust. Deep
weathering under humid tropical conditions is necessary for the formation of
both ore types, and aluminum ores are widely distributed in parts of the world
having such climates now, or in the Tertiary, but production of lateritic Ni is
largely confined to New Caledonia, the Dominican Republic, and Cuba.

Part I. Aluminum

Al, unlike most metals in sedimentary ore deposits, occurs at the Earth's surface
in only a single oxidation state, Al^{3+}. Consequently, it is not subject to the redox
cycles that play such an important part in the geochemistry of Fe, Cu, and U.
Instead, Al is concentrated almost by default, under conditions where all other
elements are removed: in the humid tropics in soils with good drainage.

Of the metals considered in this book, Al is the most abundant in the Earth's
crust. As a consequence, it is present in appreciable quantities in almost all rock
types (Table 4-1). Only peridotite and orthoquartzite are unlikely parent rocks
for bauxite; only nepheline syenite is distinctly favored. Note also the vanishingly
small concentration of Al in natural waters, reflecting its insolubility under most
conditions.

Table 4-1. Abundance of Al in common rocks and natural waters.

	$Al_2O_3(\%)$	$Al_2O_3(\%)/Fe_2O_3$
Igneous Rocks		
peridotite	4.0	0.3
basalt	14.1	1.1
andesite	18.2	1.9
granite	13.9	4.9
nepheline syenite	21.3	4.6
Sedimentary Rocks		
sandstones		
orthoquartzites	1.1	1.6
arkoses	8.7	3.5
graywackes	13.5	2.1
shales	14.7	1.8
carbonates	2.5	5.0
Natural Waters	ppm Al	
seawater	0.001	
river water	0.24	
springs in granite	0.018	

Source: Data from Wedepohl, 1969, tables 13-E-2, 13-E-6, 13-I-1, 13-K-1, 26-K-1.

There is, unfortunately, little agreement about the terminology to be used for residual soils. For instance, the term laterite has been used for such a variety of soils that its meaning has become obscured. Bauxite is the term universally employed for Al-rich laterite, with either 80 or 90 percent Al_2O_3 (water-free basis) used as the lower concentration limit. There is no analogous term for the Fe-rich end-member, however, nor is there agreement on how SiO_2 should be included. At least part of the difficulty stems from mixing chemical and mineralogical variables. Al is found as both free aluminum minerals and in kaolinite, a form undesirable for extraction, so that Al_2O_3, by itself, does not adequately describe a bauxite. Consider the compositions of tropical soils shown in Fig. 4-1. The analyses shown are from samples virtually free of clay, so that the only minerals present are free oxides and hydroxides. This being the case, it is easy to identify laterites as the low-SiO_2 base of the triangle, divided into bauxites or iron laterites based on the predominance of Al_2O_3 or Fe_2O_3. Further subdivisions can then be made based on clay content, resulting in schemes like that in Fig. 4-2. Because kaolinite is almost the only clay mineral in bauxite deposits, clay content can be approximated by using the ratio of free to combined Al_2O_3 in the sample. This quantity is sometimes available from processing data, but is readily approximated, for quartz-free samples, by

$$Al_2O_{3(free)} = Al_2O_{3(total)} - 0.85SiO_2.$$

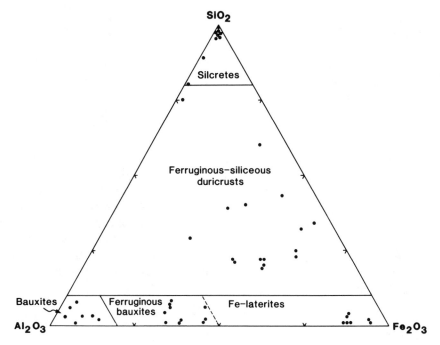

Fig. 4-1. Classification of residual soils, following Dury (1969). Analyses of some typical examples are shown, using data from Dury (1969, table 2), Goudie (1973, table 11, 14), Lelong and others (1976, table 3).

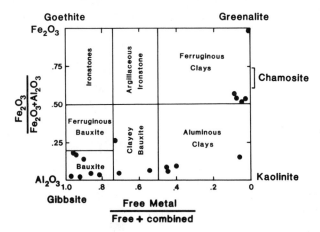

Fig. 4-2. Suggested mineralogic-chemical classification of Al and Fe ore deposits. Assumes <10 % free silica. Boundaries based on Valeton (1972, fig. 28); analyses from Arkansas bauxites (Gordon and others 1958, table 10, p. 118).

Commercial production is currently limited to ores with more than about 80 percent of the Al_2O_3 in the free state. The terminology to be used for the Al-rich part of this diagram seems well established; I have suggested an analogous set of terms for the Fe-rich portion. Because chamosite is much more variable in composition than kaolinite, it is not as easy to estimate the proportion of free metals, but a rough approximation is

$$(Al_2O_3 + Fe_2O_3)_{free} = (Al_2O_3 + Fe_2O_3)_{total} - 0.45SiO_2,$$

again on a quartz-free basis.

Within bauxites there are also several classifications used. One of the most common is based on the type of underlying rock (e.g., nepheline syenite, basalt, limestone). Deposits on limestones form an important group, the karst bauxites, that are often treated separately from lateritic bauxites (e.g., Bárdossy 1982). Alternatively, geomorphic position can be regarded as more important, resulting in terms such as plateau and slope bauxite. The most important distinction, however, is between autochthonous and allochthonous. Many bauxites appear to have been transported appreciable distances, so distinguishing these two types is the essential first step in any genetic interpretation.

Mineralogy

Aluminum is currently obtained only from simple oxides and hydroxides. These form two analogous series designated α and γ (Table 4-2). The α series minerals have an open structure, with oxygens in adjacent layers opposite one another; γ minerals have a more closely packed structure, with oxygens in one layer matched to vacant positions in the next (Hsu 1977, p. 99-106). Nordstrandite has an intermediate structure made up of alternating gibbsite and bayerite layers.

Table 4-2. Common aluminum minerals.

Name	Formula	Molar Volume (cm^3/mole)
α series		
bayerite	$Al(OH)_3$	31.15
diaspore	$AlOOH$	17.76
corundum	Al_2O_3	25.37
γ series		
gibbsite	$Al(OH)_3$	32.00
boehmite	$AlOOH$	19.53
(synthetic)	Al_2O_3	27.8
Others		
nordstrandite	$Al(OH)_3$	32.28

Source: Perkins and others 1979, table 5.

Surprisingly little is known about what controls the Al mineralogy of bauxites. For instance, of the trihydrates, bayerite is almost unknown in nature (Gross and Heller 1963) yet is easily prepared in the laboratory by acidifying Na-aluminate gels. Hem and others (1973) found that gibbsite crystallizes from solutions with a pH below 6, bayerite from solutions with a pH above 7. If silica is present, prolonged aging is required before crystalline products appear. Other laboratory studies confirm the control by pH, in some cases with nordstrandite appearing at intermediate pH as an aging product of bayerite (Barnhisel and Rich 1965, Schoen and Roberson 1970). pH values higher than 7 are certainly common in natural systems, so the scarcity of bayerite is still puzzling.

Boehmite and diaspore present a similar problem. Although diaspore is common, gibbsite and boehmite make up the bulk of the minerals in bauxites, but, as yet, there is no agreement as to why particular deposits have one mineral instead of another. Valeton (1964) suggested that perhaps oxidation potential is important; that is, the sequence diaspore-boehmite-gibbsite may reflect increased oxidation. However, a survey of recent descriptions of bauxites that report both chemistry and mineralogy (Grubb 1970, Ashraf and others 1972, MacGeehan 1972, Kronberg and others 1979) shows no relationship between mineralogy and either total iron or oxidation state of the iron. A similar conclusion has been reached by Bárdossy (1982, p. 214) for karst bauxites. Several lines of evidence suggest that more intense leaching and a drier soil favor boehmite and diaspore over gibbsite. One strong argument is that most karst bauxites (except those of Jamaica) are boehmitic or diasporic, while those on other rock types tend to be gibbsitic, possibly reflecting rapid drainage in karst soils (Valeton 1972, p. 179). Also, older bauxites are commonly boehmitic, younger ones gibbsitic, with a parallel change in iron mineralogy: hematite accompanies boehmite and goethite accompanies gibbsite (see, for example, table 16 and 18 in Bárdossy 1982). The distribution of minerals in mixed boehmite-gibbsite deposits also suggests control by drainage. The Aurukan bauxite of Australia (MacGeehan 1972) is characterized by an upper boehmite zone, about 3 m thick, overlying 6 m of gibbsite. MacGeehan hypothesized that the boundary marks the former position of the water table, boehmite having formed in the zone of leaching and gibbsite in the zone of saturation. Iron, as hematite, is largely confined to the upper, boehmitic zone.

If leaching and aging control the distribution of boehmite and gibbsite, what about diaspore? For a time it was believed that diaspore could only form under metamorphic or hydrothermal conditions (Ervin and Osborn 1951), but many occurrences are hard to reconcile with such an origin (Valeton 1964). Analogy with the dehydration of gibbsite to boehmite suggests that conversion to diaspore should also be favored by time and by intense leaching. Kiskyras and others (1978) have presented data from boehmitic and diasporic bauxites on karst in Greece that show a strong inverse relation between diaspore and kaolinite (Fig. 4-3). Apparently, diaspore forms only in environments from which silica has been almost completely removed. These authors believed that the rate of drainage was not the factor controlling the distribution of diaspore, because of a lack of

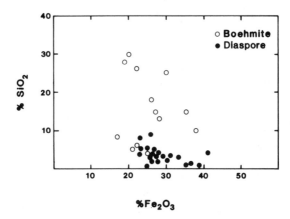

Fig. 4-3. Relationship between mineralogy and bulk chemistry in karst bauxites of Greece. Fe_2O_3, a function of oxidation state, has no effect, but SiO_2, which is equivalent to kaolinite in this case, favors boehmite (Data from Kriskyras and others 1978, table A-C.)

detailed correlation between diaspore occurrence and maximum drainage. Instead, they proposed that oxidation state was more important, with diaspore favoring the center of dolines because of lower Eh. There is no reason, however, that kaolinite should be sensitive to oxidation state. More likely, dissolved silica inhibits the transformation of gibbsite or boehmite to diaspore, and the reaction only occurs in soils from which almost all of the silica has been removed.

Other minerals in bauxite are even more poorly known. Nordstrandite is occasionally reported (Wall and others 1962, Hathaway and Schlanger 1965) but nothing is known about what controls its presence. Titanium as well as aluminum is enriched in residual soils, but not to commercial grades. The mineral form most commonly reported is anatase. Iron minerals are also abundant, usually goethite or hematite, but siderite ($FeCO_3$) is often found in the lower part of the soil profile, showing that reducing conditions were present and that iron was mobile as Fe^{2+}. Occasionally even chamosite is found, if the soil was later covered by a swamp (Nikitina and Zvyagin 1973, Bárdossy 1982, p. 210-213).

Geochemistry

Chemical analyses of bauxites show them to be enriched in Al and Ti, and sometimes in Fe, relative to the parent material (Table 4-3), but strongly depleted in more soluble elements (K, Na, Ca, Mg). The underlying rock type exerts less influence than one would expect. Bauxites on basic igneous rocks are Fe-rich, those on limestones are higher in Ca, but otherwise there are no obvious trends. Another observation is that Ti is often depleted relative to Al. Commonly Ti is regarded as being immobile during weathering, and the analyses do show an increase in TiO_2, but the decrease in Ti/Al shows that Ti is, in fact, being lost from most bauxites. Gardner (1980) reached the same conclusion from a study of Al_2O_3 and TiO_2 in soils as a function of density.

What are the geochemical controls that produce this pattern? As already mentioned, Eh has no effect on Al, so the type of Eh-pH diagrams we have used for other metals do not apply. The solubility of Al minerals is, however, strongly dependent on pH (Fig. 4-4). For instance, gibbsite is soluble at both low and high pH, with a minimum solubility in the pH range 5 to 9, the range into which most surface waters fall. Because the other elements that are common in the Earth's crust are generally more soluble under these conditions, Al accumulates in soils. In fact, the relative immobility of Al is often used in constructing equilibrium diagrams for weathering (Fig. 4-5). The number of solution variables that need to be considered can be reduced if Al is regarded as being conserved in the solids:

$$2Al(OH)_3 + 2H_4SiO_4 \rightarrow Al_2Si_2O_5(OH)_4 + 5H_2O.$$
(gibbsite)   (kaolinite)

Kaolinite and gibbsite are common minerals in tropical soils, muscovite and microcline are common in igneous rocks. The diagram shows that as alkalis such as K are removed from the soil, the primary igneous minerals are converted to kaolinite. Attack by hydrogen ion, either from CO_2 in the soil or from plant acids, is the principal agent for this transformation. If large amounts of water are present, decomposition can proceed further with leaching of silica from the soil, converting the kaolinite to gibbsite. Note that gibbsite can also form directly

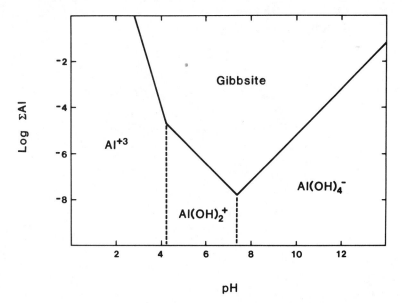

Fig. 4-4. Solubility of gibbsite, $Al(OH)_3$, as a function of pH. Because the form of Al in solution changes, there is a region of greater solubility at both high and low pH.

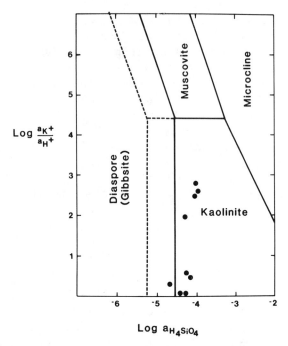

Fig. 4-5. Equilibrium diagram for some soil minerals, showing the expected solution composition in contact with each. Solid lines are for diaspore, dashed lines for gibbsite, which is metastable. The dots show observed compositions of soil waters from bauxites. (Data from Patterson and Roberson 1961, table 1; Hill and Eglington 1961, table 1.)

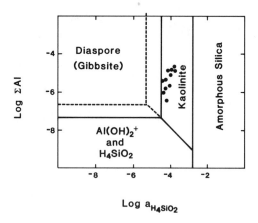

Fig. 4-6. Stability fields for several soil minerals at pH = 6 as a function of dissolved Al and Si, with observed water compositions. (Data from Patterson and Roberson 1961, table 1; Grubb 1970, table 3.)

from muscovite or feldspar, and it can be resilicated to kaolinite if conditions change slightly and $a_{H_4SiO_4}$ rises. The silica in both these reactions is from silicate minerals rather than from quartz. Quartz is extremely slow to react at surface temperatures, so that quartz-rich rocks seldom develop bauxites, even when equilibrium considerations predict that quartz will dissolve.

So far, we have been using gibbsite as the aluminum mineral and, indeed, it is the most common one in bauxites. Fig. 4-5 shows, however, that diaspore is the stable phase. An alternative representation, in which a field of complete dissolution is included (Fig. 4-6), shows the same effect. Also, note that natural water analyses from bauxitic soils conform more closely to the diaspore-kaolinite boundary than to gibbsite-kaolinite. Equilibrium among the aluminum hydrates is controlled by a_{H_2O} (Fig. 4-7). Note that boehmite is everywhere metastable with respect to diaspore, even though it is common in bauxites. Also, diaspore is stable relative to gibbsite for any a_{H_2O} less than about 5. Because the maximum value for this quantity is 1, diaspore is the only stable phase at low temperatures. Kittrick (1969) came to the same conclusion and suggested that there is an aging sequence

amorphous $Al(OH)_3 \rightarrow$ gibbsite $\rightarrow$ boehmite $\rightarrow$ diaspore

seen in the rock record (Fig. 4-8): Cenozoic bauxites are mostly gibbsitic, Mesozoic ones boehmitic, while most Paleozoic high-alumina clays contain diaspore (Bridge 1952). Although we do not yet understand all of the factors that favor the appearance of gibbsite rather than direct formation of diaspore, metastable intermediates appear to be the rule in low-temperature systems (Paces 1978). One factor that has recently been identified is the concentration of carbonate ion (Bárdossy and White 1979). In laboratory studies, the rate of crystallization of

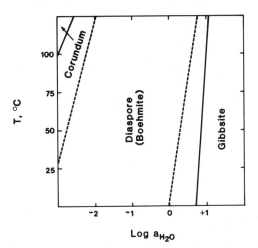

Fig. 4-7. Diaspore is the only stable Al mineral at low temperatures and reasonable activities of water. The dashed lines show metastable equilibria involving boehmite.

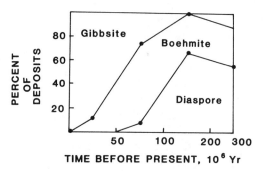

Fig. 4-8. Mineralogy of karst bauxite changes with time in the direction of increasing dehydration. (Data from Bárdossy and others 1978, table 1.)

$Al(OH)_3$ is much slower in the presence of CO_3^{2-}. Further, the grain size of gibbsite is considerably smaller in bauxites developed on carbonate rocks compared with other rock types. If this inhibition proved to be stronger for gibbsite than for boehmite, it would explain the association of boehmitic bauxites with limestones.

Petrography

Bauxites exhibit relict, massive, or concretionary textures. In relict textures, the bauxite preserves the texture of the parent rock. Although not always evident in hand sample, this type is usually quite obvious in outcrops. Massive, featureless bauxites also occur, but the concretionary type is the most common. Almost all bauxites, somewhere in their profile, have a zone of concretionary bodies. If less than 2 mm in diameter, they are termed ooliths, if greater, pisoliths. The distribution of sizes is rarely reported, but in the Arkansas deposits, ooliths are generally less than 1 mm, while pisoliths are 3-20 mm, with few particles of intermediate size (Gordon and others 1958, p. 76). Packing is another textural property that is seldom studied. Commonly, the ooliths and pisoliths "float" in a structureless matrix and are very poorly sorted; in other cases there is close packing and better sorting. In the first case one is tempted to infer that the bauxite formed in place, in the second that it has been transported. Occasionally transported bauxites have discernible sedimentary structures such as cross-bedding, but generally such evidence is rare.

On a microscale, the ooliths and pisoliths are seen to comprise alternating Fe-rich and Fe-poor laminae, the number and thickness of which are variable. Scanning electron microscopy shows that the laminae are made up of tangentially oriented particles, apparantly the result of crystallization from solution (Bhattacharyya and Kakimoto 1982). Certain diagenetic features are also revealed in thin section. For example, the Aurukun bauxite of Australia (MacGeehan 1972) underwent two stages of aluminum enrichment that can be followed petrographically. In the first, boehmite pisoliths formed above the water table, gibbsite below. Both types of pisoliths contain syneresis cracks filled with gibbsite. In the second stage, associated with a drop in the water table, an envelope of boehmite or gibbsite formed around pre-existing pisoliths, cementing them into pairs or triplets.

Kaolinization is another common diagenetic feature seen in thin section. Most bauxite deposits contain considerable kaolinite or halloysite, and there has been much debate over whether such minerals represent an intermediate stage in the conversion of feldspar to bauxite, or are secondary products formed by later introduction of dissolved silica. Bates (1960) illustrated halloysite as a precursor to gibbsite in soils of Hawaii. In contrast, Gordon and others (1958, plates 12, 13) presented a series of photomicrographs from Arkansas bauxites that show resilication of gibbsite to kaolinite, as evidenced by kaolinite veins and pore fillings. They concluded that, although some kaolinite must have formed directly from the parent igneous rock (a nepheline syenite), most is of this secondary type. Thus, both mechanisms of kaolinite formation seem to apply, that is, the kaolinite $\rightarrow$ gibbsite reaction is reversible.

Most undisturbed bauxites have a vertical profile of petrographic features (Mindszenty 1978). At the surface is a zone of leaching, characterized by a relative enrichment of Al through removal of the other constituents. This is underlain by a zone of accumulation, where there is an absolute enrichment of Al or Fe, followed, finally, by a zone of decomposed parent rock. In the zone of leaching, thin-section petrography often shows direct transformation of feldspars to gibbsite, with good preservation of the original textures. Lower in the profile, absolute enrichment is shown by pore-filling gibbsite cements. Decomposition of the parent rock is accompanied by microbrecciation structures. These three types of structures are often superimposed, probably because of fluctuations in the ground water table. An additional class of structures is what we might call "aging structures," which develop with time in these other zones. Pisoliths are the best example. They seem to form through recrystallization of the minerals, accompanied by separation of Fe and Al into discrete layers, by a process that is not well understood.

Another petrographic approach, which is just beginning to be used and shows great promise, is scanning electron microscopy (SEM). Bárdossy and others (1978) investigated a number of deposits and found that karstic bauxites are characterized by a much smaller grain size than other types, and that grain size increases with elevation in lateritic bauxites. Karst bauxites also show changing textures with increasing compaction, going from open stacking, to microporous, to compact, and, finally, to a recrystallization texture referred to as "crystalline-webby." An interesting extension of this study would be to examine bauxites of different ages to see if analogous changes can be seen with time. The transformation of feldspar to gibbsite, both directly and via kaolinite, can also be observed with the SEM (Keller 1979).

Vertical Sequence

The vertical sequence encountered in bauxite deposits depends on the type of parent rock—principally its iron content—and such factors as slope and drainage. Confining ourselves for the moment to *in situ* lateritic bauxites, an Fe-rich and an Fe-poor profile can be distinguished (Fig. 4-9). In the Fe-poor case, Al enrichment is mostly relative, by removal of other substances with many of the structures of the parent rock preserved; in the Fe-rich case more of the enrichment is absolute (Lelong and others 1976, p. 115). Fe is left behind as a duricrust, while Al is moved lower in the profile. In both cases, it is common to find truncations, repetitions, and superpositions. For example, an organic-rich A horizon should normally be present, but it is often removed by erosion exposing the underlying Fe-Al accumulation to desiccation and lithification. A second cycle of bauxitization may then begin, forming another bauxite layer deeper in the soil (Valeton 1972, fig. 47). Another type of disruption of the ideal profile is silicification of the Al minerals to kaolinite, usually in the surface layer or at the top of the parent rock. Of course, these vertical zones do not persist indefinitely in the horizontal direction. Instead, there is a set of lateral facies changes, usually related to the nature of the slope, and similar to the vertical changes (Valeton 1972, fig. 35).

For transported bauxites, the situation is more complex. First, let us consider the Arkansas deposits where the distinction between the *in situ* and transported bauxites is clear. During the time these deposits were forming (Paleocene to Early Eocene), hills of nepheline syenite stood up as positive areas. These were gradually overlapped by marginal marine and non-marine strata that incorporated considerable bauxitic debris from soils on the syenite hills (Fig. 4-10). For the *in situ* deposits (Type 1), the vertical sequence is much like that shown in Fig. 4-9, except that a zone of secondary kaolinite appears at the syenite-bauxite contact (Gordon and others 1958, fig. 36). The transported bauxites have been divided into three types. The first (Type 2) deposits formed as fans, deposited by mass-wasting rather than by streams, along the lower slopes of the syenite

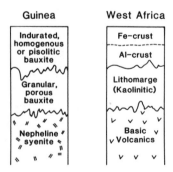

Fig. 4-9. Bauxitic soil profiles in Fe-poor and Fe-rich parent rocks in the same climate. In the second case, a resistant Fe cap develops at the surface while Al accumulates below.

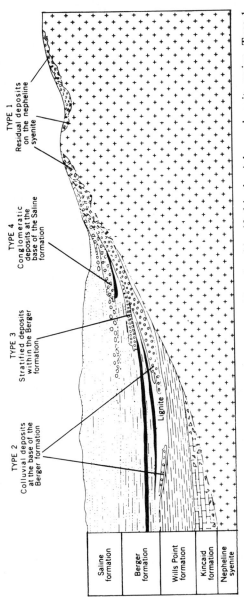

Fig. 4-10. Diagrammatic section showing the four types of bauxites identified in the Arkansas bauxite region. Type 1 is an *in situ* deposit, while 2-4 are all allochthonous (Gordon and others 1958, fig. 33).

hills and in places extending onto the fringing sediments. They are believed to have originally comprised a mixture of bauxite and kaolinite that was subsequently leached to pure bauxite on valley sides, but remained largely kaolinitic in valley centers. A typical vertical profile shows a "fragmental underclay" at the base, becoming progressively richer in gibbsite, until a zone of pisolitic bauxite is reached. This is overlain by a "siliceous hardcap" and then a kaolinitic clay. The "fragmental underclay" consists of angular to rounded fragments of kaolinite or bauxite, up to 10 cm in length, set in a matrix of kaolinite, with some siderite. The "siliceous hardcap" comprises soft gibbsitic pisoliths in a hard kaolinitic matrix, and is believed to have resulted from resilicification under swampy conditions. Type 3 and 4 deposits do not show evidence of the extensive post-depositional bauxitization and resilicification found in Type 2. They consist of either moderately sorted, closely packed pisoliths (Type 3), or large boulders of bauxite in a matrix rich in clay and sand (Type 4). Neither type has a discernible internal zonation like Type 1 or 2, and they seem to have been deposited by local streams as purely clastic accumulations with little chemical modification.

Distinguishing transported and *in situ* bauxites is especially difficult in deposits overlying carbonate rocks, because of the absence of a transitional layer of slightly weathered parent rock. Further, the aluminum content of the host rock is usually so low that it is hard to envision how significant amounts of bauxite could form in place by the alteration of the insoluble residue in limestone or dolomite. There are four ways in which such deposits can be formed:

1. Autochthonous—strict *in situ* development from the underlying host rock.

2. Parautochthonous—from weathering of the host rock, but with local transport into low-lying areas, especially dolines on karst topography.

3. Allochthonous—derived from some more or less remote parent rock different from the host.

4. Crypto-autochthonous—derived from weathering of a parent material (e.g., volcanic ash) that is now completely altered, so that the bauxite rests conformably on an unrelated host rock.

The difficulties in distinguishing these types are illustrated by the bauxites of Jamaica. There is no transition zone between the limestone and the bauxite, which has led to sharply divided opinions as to an autochthonous (Hose 1963, Ahmad and others 1966, Sinclair 1966 and 1967, Smith 1970) or an allochthonous origin (Zans 1959, Burns 1961, Kelly 1961, Roch 1966). Surprisingly, none of those proposing derivation of the bauxite from the limestone seem to have considered a parautochthonous origin. The proponents of the allochthonous view have not argued that the bauxite itself has been transported, rather that it was formed by *in situ* weathering of clastic debris eroded from volcanic rocks at higher elevations. Because no trace of these clastics remains, this model is better classed as crypto-autochthonous. Comer (1974) has proposed a new crypto-autochthonous model that overcomes some of the objections to the fluvial transport required by the earlier ones. He suggested that the limestone surface was originally covered by a blanket of volcanic ash as much as 6 m thick, which is still preserved

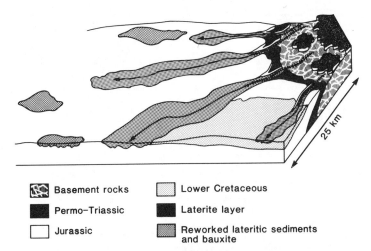

Basement rocks

Permo–Triassic

Jurassic

Lower Cretaceous

Laterite layer

Reworked lateritic sediments and bauxite

Fig. 4-11. Allochthonous bauxite deposits of France. The detrital laterites and bauxites are preserved in synclinal lows in the pre-Middle Cretaceous surface (after Veleton 1972, fig. 66).

in a few places. This layer then provided the Al for the bauxite. The presence of a karst surface was crucial, however, because it provided for intense vertical leaching with little surface erosion—an ideal condition for the formation of bauxite or laterite (Clarke 1966; Comer 1974).

Bauxites on karst in southern Europe are in some cases clearly allochthonous. For example, Nicolas (1968) has argued that many of the deposits of southern France are detrital accumulations derived from the weathering of lateritic crusts, as evidenced by sedimentary structures, presence of fossils, and a transitional contact with the overlying sediments, contrasted with a sharp discontinuity with the underlying limestone. In some cases, considerable post-depositional alteration has occurred, most commonly recrystallization of the Fe and Al minerals and silicification. The elongate, channel-like form of these deposits also argues for an allochthonous origin (Fig. 4-11).

Theories of Origin

The general conditions under which bauxites form are not in doubt: a humid tropical climate with sufficient drainage for Na, K, Ca, and Mg to be removed. It is not quite so certain, however, how the Al is separated from the Si and Fe. The Si contained in aluminosilicates, such as feldspars or clays, seems to pass into solution relatively readily, and so is washed from the soil under the above conditions. In contrast, quartz dissolves extremely slowly, and, although thermodynamics predicts quartz should be more soluble than gibbsite for pH greater than 4 (Fig. 4-12), quartz persists in tropical soils. As a consequence, bauxites form preferentially on rocks that are low in quartz such as syenites, basalts, and

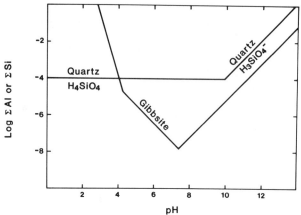

Fig. 4-12. Quartz is theoretically more soluble than gibbsite for pH greater than 4, but because of its slow rate of dissolution it tends to persist.

limestones (Fig. 4-13). Furthermore, they are favored by long time periods of intense chemical weathering but little erosion.

Separation of Al from Fe must depend on some other factor, because, under oxidizing conditions, gibbsite is more soluble than goethite at all pH values (Fig. 4-14), and both phases react quickly. For low-iron bauxites to form, it therefore seems essential for either the parent rock to have been low in iron, or for conditions to have prevailed under which either Fe or Al can be mobilized. Fe is depleted in the Arkansas bauxites, from an Fe_2O_3/Al_2O_3 ratio of 0.18 in the syenite to 0.03 in the bauxite (Table 4-3), but it can be seen that Fe was originally low. In comparison, the basalts on which the Hawaiian bauxites formed are much richer in Fe, with Fe_2O_3/Al_2O_3 of about 1.0, which increases in the bauxites to 1.7, resulting in a high-iron bauxite. Low-iron bauxites can form from such

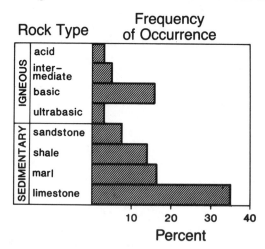

Fig. 4-13. Bauxites can form on almost any parent rock, but show a preference for those low in quartz such as basalt and limestone (after Sklijanov 1978, fig. 1).

Table 4-3. Bulk chemical composition of some typical bauxites and underlying rocks.

	Al_2O_3	SiO_2	Fe_2O_3	FeO	TiO_2	H_2O	MgO	CaO	Na_2O	K_2O	TiO_2/Al_2O_3
Arkansas[a]											
Bauxite	59.9	4.1	1.3	0.4	2.2	31.2	—	—	—	—	0.037
Bauxitic clay	44.3	33.1	1.4	0.3	1.7	18.5	—	—	—	—	0.038
Nepheline/syenite	19.2	58.7	1.7	1.5	0.9	1.5	0.7	1.4	7.5	6.3	0.047
Hawaii[b]											
Bauxite	25.6	5.7	40.2	0.4	5.6	18.6	0.7	(tr)	0.04	0.12	0.22
Basalt	14.8	45.0	4.3	9.3	1.9	3.1	12.0	9.7	1.8	0.62	0.13
India[c]											
Bauxite	53.5	0.8	10.1	—	7.3	28.2	—	—	—	—	0.14
Basalt	14.4	48.1	15.6	—	2.7	19.7	—	—	—	—	0.19
Bauxite	62.4	2.7	1.2	—	2.5	31.2	(tr)	(tr)	—	—	0.04
Weathered tuff	14.1	49.9	9.8	—	3.0	23.0	(tr)	(tr)	—	—	0.21
Bauxite	58.2	1.7	4.8	—	5.0	30.4	0.4	—	—	—	0.09
Weathered shale	39.4	42.3	2.0	—	1.7	15.0	0.5	—	—	—	0.04
Bauxite	64.1	5.5	0.2	—	5.1	22.0	—	2.8	—	—	0.08
Clay	13.6	25.7	37.2	—	4.6	11.5	—	7.4	—	—	0.34
Weathered limestone	5.6	2.7	61.0	—	2.7	12.8	—	15.2	—	—	0.48

Sources: [a]Gordon and others 1958, table 6, 10.
[b]Valeton 1972, table 16.
[c]Sahasrabudhe 1978.

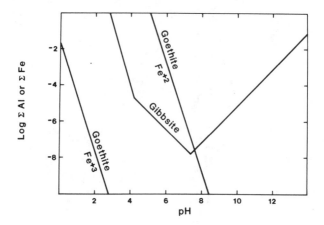

Fig. 4-14. Gibbsite is more soluble than goethite at all pH's under oxidizing conditions (Fe as Fe^{3+}), but below a pH of 8, goethite is the soluble phase under reducing conditions (Fe as Fe^{2+}).

parent material by leaching of the Al and reprecipitation lower in the profile. Many bauxites show a cap of iron-rich laterite overlying the bauxite (Fig. 4-9), suggesting absolute enrichment of the Al at depth and relative enrichment of Fe at the surface. Relative enrichment of Al, by removal of the Fe from lateritic soils, can occur under reducing conditions (Petersen 1971; Norton 1973), and the common presence of plant debris, either within or immediately overlying the bauxite, and Fe^{2+} minerals such as siderite and chamosite show that reduction can occur. A well-documented example is the deposits of the northern Cape York Peninsula of Australia (A.H.White 1976). Here, bauxite formed on a gently undulating surface, and was subsequently covered by sandy sediments. In depressions, this cover was a brown, lignitic sand, but over the rest of the surface, clean white sands were deposited. The original bauxite had an iron oxide content of 15 to 20 percent, but beneath the lignitic sands the content is less than 2 percent (Fig. 4-15). Norton (1973) pointed out that in such cases of relative enrichment owing to removal of Fe, there should also be a depletion in Mn, Co and Ni. What should happen in cases of absolute enrichment of Al is not clear, however, and the available field evidence (e.g., Dennen and Norton 1977) does not show a distinctive difference in trace elements between these two processes.

Summary

Al ores are soils that result from extreme leaching under humid tropical conditions. They require high drainage rates and intense chemical weathering, combined with little physical erosion. Parent rock is of lesser importance except that high-iron parents, such as basalt, tend to produce ferruginous bauxites. Whether Fe or Al is enriched in a given profile can be explained using Eh-pH relations.

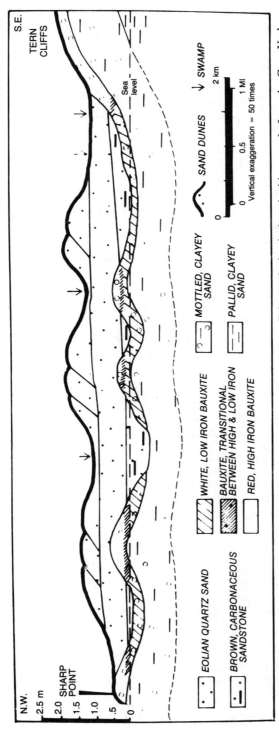

Fig. 4-15. Low-iron bauxites form under swampy, reducing conditions where Fe^{2+} can be leached, as in this example from the Cape York peninsula of Australia (A. H. White 1976, fig. 2).

The most vexing problem is the distribution of aluminum minerals. There seems to be a rough age zonation gibbsite $\rightarrow$ boehmite $\rightarrow$ diaspore that suggests kinetic control of mineralogy. More laboratory and field observations are needed to quantify this behavior. Another area needing attention is the microchemistry of bauxitization. For instance, what is the composition of zones of absolute enrichment of Al compared with zones of relative enrichment? Along the same line, how do the different layers in the pisoliths form? These and similar questions require careful work with the petrographic microscope and electron microprobe.

PART II. NICKEL

Nickel, like aluminum, is found in only a single oxidation state (Ni^{2+}) at the Earth's surface. Unlike Al, it forms several stable sulfides, and these are the common ore minerals in the large magmatic deposits, such as Sudbury. Lateritic weathering is also capable of concentrating nickel from ultrabasics, usually serpentinites, thereby forming important, but lower-grade deposits. The necessary combination of lateritic weathering and a nickel-rich parent rock makes commercial deposits of this type uncommon. As with bauxites, both autochthonous and allochthonous varieties are known, but the former greatly predominate.

Ni is usually present in a dispersed state in the Earth's crust because of its ready substitution for Fe and Mg in silicates. Consideration of abundances (Table 4-4) shows that sedimentary nickel accumulations can only occur in association with the pre-existing Ni concentration of ultrabasic igneous rocks. An exception

Table 4-4. Abundance of Ni in common rocks and natural waters.

	ppm Ni
Igneous Rocks	
ultramafics	1450
basalts	130
andesites	18
granites	10
Sediments and Seawater	
Pacific Mn nodules	3120
Pacific deep-sea sediments ($CaCO_3$-free)	300
near-shore clays	40
seawater	0.0005
Sedimentary Rocks	
shales	70
black shales	50
graywackes	40
limestone	5

Source: Data from Wedepohl 1969, table 28-E-1, 28-I-1, 28-K-1 to 3.

is the marked concentration seen in deep-sea manganese nodules, which may be an important future resource for Cu and Co as well as Ni (see Chapter 5).

Mineralogy

The most common minerals in lateritic nickel deposits are goethite, serpentines, talc, and smectites (Fig. 4-16). These may contain Ni substituting for Fe or Mg; if the proportion exceeds a mole fraction of 0.5, a separate name is given, otherwise the phase is referred to as Ni-goethite, Ni-talc, etc. (Table 4-5). Because of poor crystallinity, the Ni-silicates are often hard to identify, and the general term garnierite is then used. Depending on whether Ni-goethite or Ni-silicate predominates, nickeliferous laterites are divided into an oxide and a silicate (or

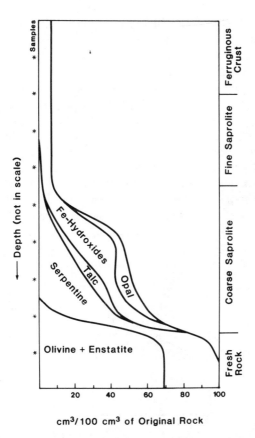

cm³/100 cm³ of Original Rock

Fig. 4-16. Typical mineralogic profile of a Ni laterite. In this example from New Caledonia, the profile is about 30 m thick, with the "fine saprolite" being the thickest zone. Ni contents of the sample intervals marked are given in Table 4-5 (after Lelong and others 1976, fig. 12).

Table 4-5. Terminology for Mg-Ni silicates.

Mg end-member	Ni end-member
1:1 minerals (serpentines)	
chrysotile	pecoraite
lizardite	nepouite
Intermediate	
kerolite	pimelite
2:1 minerals	
talc	willemsite
chlorite	nimite
sepiolite	falcondoite

Source: After Brindley 1978, Springer 1974.

saprolite) zone. Both are normally present in autochthonous deposits, but commercial mineralization is usually confined to one or the other.

The oxide zone is made up predominantly of goethite. Hematite and sometimes maghemite are found in smaller amounts towards the top of the profile (Elias and others 1981, fig. 2). The goethite usually has some Al in solid solution, the amount of which decreases downwards in the soil profile as silicate minerals begin to compete for aluminum (Fig. 4-17). Ni is found both in association with the goethite and in discrete cryptocrystalline masses of Mn-Co oxide termed "asbolite" (Chukrov and others 1982). Schellmann (1978) has presented evidence that the bulk of this Ni is in solid solution with Fe in the goethite. He used the output from two spectrometers of a microprobe, recorded on an X-Y plotter, to examine the relationship of Fe to Ni or Cr, and found evidence for the presence of two phases, one of which showed a close correlation of Ni and Fe, and one which was essentially free of Ni. He identified the first as goethite, the second

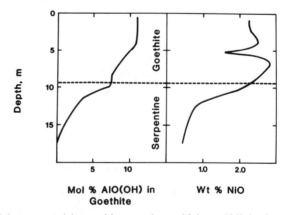

Fig. 4-17. Al incorporated in goethite may be as high as 10% in the oxide zone, but decreases in the saprolite zone. In this deposit, note that mineralization is dominantly in the oxide zone. (Data of Zeissink 1969, tables 1, 2.)

as maghemite, but in the discussion of the paper it was suggested that ferrihydrite might also be present and, because of its high sorption capacity, might be a more likely repository for the Ni (Kuhnel and others 1978, p. 202). Schellmann (1978) also presented several interesting scanning electron photomicrographs that show two kinds of particles, needles and platelets; the first are probably goethite, while the second may be ferrihydrite (Kuhnel, in discussion of Schellman 1978, p. 282). Cr proved to also be associated with goethite, but more loosely held, probably by adsorption; Co was associated exclusively with Mn oxides. This combination reaches commercial concentrations in the Kalgoorlie area of Australia, where a thin Mn oxide horizon at the base of the oxide zone contains as much as 2 percent Co (Elias and others 1981, p. 1779). The association of Co with manganese phases and Ni with iron is interesting, because it is the reverse of the pattern found in deep-sea manganese nodules (Chapter 5). The cause of this partitioning is not known, but may be related to the oxidation state of the Co, whether $2+$ or $3+$, or to the particular Mn minerals present, todorokite favoring Ni and γ-MnO_2 favoring Co (Cronan 1980, p. 132-134).

The silicate zone usually preserves textures of the parent rock, hence the synonym "saprolite facies." In many profiles, it is separated from the oxide zone by a relatively thin layer of smectitic clay with secondary silica, either opal or chalcedony (Elias and others 1981, fig. 2). Below this clay zone, Ni-bearing minerals are mostly garnierites. Both talc and serpentine-like forms are present (Faust 1966, Brindley and Hang 1973) and in all proportions (Kato 1961, fig. 1). Phases with Ni>Mg are rare, so that most ore consists of nickelian varieties of the common hydrous Mg silicates (Table 4-6). Among the possible 7Å types (i.e., serpentines), both lizardite and chrysotile are abundant, but antigorite is

Table 4-6. Microprobe analyses of minerals in nickel laterites.

	SiO_2	Al_2O_3	FeO	MgO	NiO	CoO	MnO_2	Source
Oxides								
Asbolite	0.36	0.72	0.62	3.13	7.34	8.85	58.96	1
	0.19	0.19	0.18	0.61	11.61	1.78	61.35	1
Limonite	0.8	15.0	14.3	0.5	3.4	7.4	33.6	4
	3.20	6.09	69.8	0.46	2.85	(n.d.)	0.06	3
Silicates								
Serpentine	41.7	0.07	0.72	36.0	5.74	—	—	2
	42.7	—	6.7	41.6	0.3	—	—	5
Talc	53.8	(n.d.)	(n.d.)	17.5	19.3	—	—	2
	42.3	—	0.08	15.8	10.7	—	—	6
Sepiolite	56.9	—	0.04	17.3	17.3	—	—	6
Parent Olivine	40.8	8.56	0.04	50.9	0.45	0.05	0.19	3

1. Kalgoorlie, Australia (Elias and others 1981, table 1)
2. Brazil (Esson and Carlos 1978, table 3)
3. New Caledonia (Besset and Coudray 1978, table 2)
4. Indonesia (Golightly 1979, table 2)
5. New Caledonia (Troly and others 1979, table 2)
6. Urals (talc), Ridley, Oregon (sepiolite) (Springer 1974, table 5)

seldom found (Troly and others 1979, p. 99). 10Å phases are represented by talc. A possible intermediate is the kerolite-pimelite series, but there is some inconsistency in the way these names are applied. Brindley (1978) uses kerolite to denote a hydrous variety of talc. Others (Kostov 1968, p. 375; Springer 1974, p. 387) use this term for a mixture of serpentine and talc. Pimelite is similarly ambiguous, plus it has sometimes been reported to have swelling properties, and thus to be a Ni-smectite. Brindley (1978, p. 239) cites evidence that the observed swelling behavior is caused by a small proportion of expandable layers in a dominantly non-swelling 2:1 structure. In silica-poor environments, such as those found in karst Ni deposits, and soils developed on Ni sulfide ores, a Ni pyroaurite, takovite, is found (Bish and Brindley 1977, Bish 1978). It has the formula $Ni_6Al_2(OH)_{16}CO_3(H_2O)_4$.

Geochemistry

Nickel laterites are made up almost entirely of Si, Fe, Mg, Mn, and Ni (Fig. 4-18; Table 4-7). Notice that nickel is only somewhat enriched compared with the parent ultrabasic rock. The increase in percent Ni is between 2 and 6 times, most of which is a relative enrichment caused by the removal of Mg from the profile. For example, if NiO is compared to less soluble substances such as Al_2O_3 and Cr_2O_3, there is actually a depletion of NiO in the top two zones, compared with the ratio in the parent rock, whereas the coarse saprolite zone at the base of the profile shows somewhat higher ratios than the parent rock. Thus, just as with Al, there is a strong relative enrichment, combined with some absolute enrichment lower in the profile.

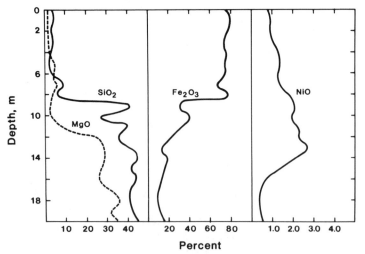

Fig. 4-18. Chemical profile of Ni-laterite at Pulot, Philippines. In this case the Ni mineralization is in the saprolite zone. (Data of Ogura 1977, table 2.)

Table 4-7. Typical chemical profile for nickel laterite of New Caledonia. The horizons correspond to those shown in Fig. 4-16.

	Parent Rock	Composition (%)		
		Coarse Saprolite	Fine Saprolite	Ferruginous Crust
Thickness		0.2–6 m	10–40 m	2–10 m
SiO_2	38	33	1.2	0.5
Al_2O_3	0.4	2	4.5	5
Fe_2O_3	3.5	17	72	74
FeO	5	2	1	0.3
CaO	0.1	—	—	—
MgO	41	29	0.9	0.5
Cr_2O_3	0.4	0.8	4.0	5.5
MnO_2	0.14	0.29	1.0	0.5
NiO	0.40	2.5	1.0	0.4
CoO	0.02	0.08	0.2	0.07
Ignition Loss	10.5	13.0	13.7	13.5
Density	2.8	1.6	0.9	—
NiO/Al_2O_3	1.0	1.25	0.22	0.08
NiO/Cr_2O_3	1.0	3.1	0.25	0.07
NiO/MnO_2	2.9	8.6	1.0	0.80

Source: After Lelong and others, 1976, table 5.

Because of a lack of thermodynamic data for Ni-silicates, it is not possible to construct precise Eh-pH diagrams for Ni-laterites. Instead, we shall use some approximations to try to develop a general picture and pose some problems for further work. Consideration of only the components Ni-O-S shows that Ni is rather soluble at high Eh, but can be immobilized as the sulfide (Fig. 4-19). One would expect, then, that its behavior would be much like that of Cu, resulting in the formation of supergene sulfide deposits. Although supergene enrichment does occur in Ni-sulfide ores (Watmuff 1974, Nickel and others 1977), it does not produce significant new orebodies, and is a different process from that responsible for the Ni-laterite deposits. Much has been written lately about alteration of Ni-sulfides (e.g., Nickel and others 1974, Thornber and Wildman 1979), especially as it relates to exploration for Ni-sulfide ore bodies (Moeskops 1977, Smith 1977), to which the interested reader is referred.

If Ni is soluble under oxidizing conditions, how is it that it is retained in the oxide zone in so many cases? As discussed above, Mn oxides in the laterites have a strong affinity for Ni—in deep-sea manganese nodules, Ni is favored over Mg by two or three orders of magnitude (Usui 1979, p. 412)—but most of the Ni is in solid solution in iron oxyhydroxides. Unfortunately, we have no thermodynamic data for these mixed oxides, and so cannot accurately represent them on equilibrium diagrams, but an idea of the geometry of such diagrams can be obtained by using trevorite ($NiFe_2O_4$), the nickel analogue of magnetite

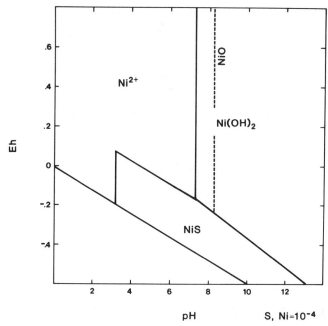

Fig. 4-19. Eh-pH diagram for Ni oxides and sulfides. The dashed line is for metastable NiO equilibrium. The large sulfide field at low Eh plus dominance of Ni^{2+} at high Eh suggests that Ni should be subject to supergene enrichment similar to copper, if sulfides are present.

(Fig. 4-20). This representation suggests that at moderate to low pH, nickel can be held in mixed Fe-Ni phases, but that at pH values above neutrality, Ni-silicates are favored, a pattern that matches the pH distribution of soil waters from Ni-laterites, which typically increase from 5 to as much as 8.5 downwards from the top of the oxide zone to the base of the saprolite (Golightly 1981, p. 723; Burger 1979, p. 31). Note that, even bound in an Fe-rich phase, Ni has a modest solubility, which decreases dramatically in the region of silicate stability. Therefore, one would expect Ni to be slowly leached from the oxide zone and reprecipitated in the saprolite zone, as was inferred to be the case based on bulk chemistry of the zones (Table 4-5).

In the saprolite zone, the Ni is held in solid solution in Mg-silicates. It has not yet been determined whether the introduction of the Ni involves a dissolution-reprecipitation of the silicates, or is accomplished via ion-exchange without alteration of the framework. Nor is it known what conditions favor Ni-talc over Ni-serpentine. The replacement of Mg by Ni is favored by the presence of unfilled d orbitals in Ni^{2+}, which give rise to crystal field effects not encountered in Mg^{2+} (Burns 1970). These lead to an appreciable increase in the stability of ionic bonds for octahedral coordination (the crystal field stablilization energy) and, thus, to much lower solubilities for Ni-silicates than for Mg-silicates (Fig. 4-21). Crystal field theory also predicts that Ni^{2+} should be slower to react than most other transition elements (e.g., Fe^{2+}) in substitution reactions (Burns 1970, p. 162-165), but there is no evidence that slow reactions inhibit the enrichment

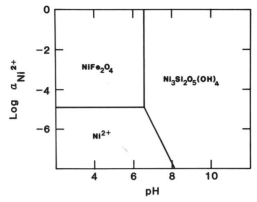

Fig. 4-20. Possible phase relations of Ni oxides, silicates, and dissolved Ni^{2+}. Ni is less soluble as the oxide than shown in Fig. 4-19 because of the presence of Fe. Note also the predominance of silicates over oxides at pH greater than approximately 6 ($a_{Fe^{2+}} = 10^{-6}$, $a_{H_4SiO_4} = 10^{-3}$, magnetite present except in trevorite field.)

of Ni in Ni-laterites. For the question of talc vs. serpentine, it is likely that the amount of serpentine in the parent rock exerts an influence. For instance, Golightly (1979a, p. 49) reported that saprolites on unserpentinized ultramafics of Indonesia have garnierites with a high proportion of talc-like layers. Alternatively, the most important factor could be the amount of dissolved silica in the pore water (Fig. 4-22). More work is needed on the field relationships of these two minerals. A final uncertainty is the nature of Ni-Mg solid solution in garnierites. The equilibrium diagrams presented here use pure phases, and for the Ni species the free energies have been approximated by extrapolation from higher temperatures (Golightly 1981, p. 722), so it is not known how well they represent the true situation for this system. Nor have I attempted to represent intermediate compositions. Furthermore, Schellmann (1982, p. 630) has shown that there is not

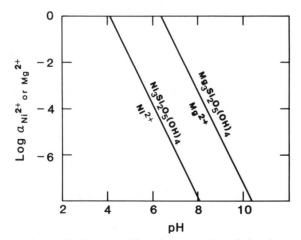

Fig. 4-21. At a given pH, far more Mg^{2+} is present in solution in equilibrium with chrysotile than Ni^{2+} in equilibrium with pecoraite (assumes $a_{H_4SiO_4} = 10^{-3}$).

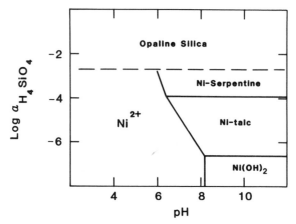

Fig. 4-22. The stability of Ni silicates is enhanced by increased pH; higher silica in solution should favor serpentine minerals over talc, but mineralogy of the parent ultramafic may also be important ($a_{Ni^{2+}} = 10^{-4}$).

a simple replacement of Mg by Ni: more Mg is lost than is compensated by the Ni gain. At present, the best we can do is try to show the general trends.

Petrography

Perhaps because of the fine grain size of the minerals, nickel laterites have not received much petrographic study. Some interesting transmission electron microscopy has been done (e.g., Zeissink 1969, Brindley 1978), but largely in an effort to identify the minerals present; the scanning electron microscope study of goethites by Schellmann (1978), discussed above, is a valuable contribution that could profitably be imitated. From thin-sections, Troly and others (1979) illustrated a sequence of alteration for the New Caledonia deposits:

1. unaltered harzburgite;
2. limonite appearing along cleavages of olivine;
3. all olivine pseudomorphed by limonite or, in some cases, saponite;
4. limonite staining of serpentine; garnierite and chalcedony deposited in cavities;
5. complete transformation of olivine and serpentine to limonite; some corroded orthopyroxenes may remain.

Several petrographic studies of chromite in nickel deposits, and of karstic nickel laterites in general, can be found in the volume prepared by Augustithis (1981). Particularly interesting is the paper by Katsikatsos and others (1981) who document the presence of clastic chromite and Ni-chlorite in karstic bauxite and nickel laterite deposits of Greece, showing the allochthonous nature of some of these ores. The effects of metamorphism on nickel ores of this area have been illustrated by Mposkos (1981): riebeckite and stilpnomelane appear in abundance, and magnetite grows to considerable size, often around nuclei of clastic chromite.

Vertical Sequence

As we have seen, autochthonous deposits of lateritic nickel show a well-developed chemical and mineralogic profile, divided broadly into an overlying oxide and underlying silicate zone, although it should be kept in mind that these zones can show marked lateral variability attributable to fracture-controlled permeability variations (Haldemann and others 1979, fig. 7). In some profiles, a transition zone consisting of nontronite and chalcedony appears (Golightly 1981, p. 719-721). Golightly (1979b, p. 14) has suggested that retarded drainage leads to an increase in the concentration of dissolved species in the soil water, and hence to saturation with respect to smectites, a supposition supported by the association with silicification, which indicates supersaturation with respect to amorphous silica. Only at Kalgoorlie, in Western Australia, is the principal Ni mineralization found in the smectite zone, and it is not obvious what factors were responsible. The area now has an arid climate, but lateritization is thought to have occurred during a wetter period during the Late Cretaceous (Elias and others 1981, p. 1780).

Some smaller deposits of nickel at the Jurassic-Cretaceous unconformity in the Balkan Peninsula appear to be allochthonous (Maksimovic 1978, Katsikatsos and others 1981). A range of types of deposits is found at this horizon, formed from weathering of:

1. Peridotite masses thrust over limestone.
2. Clastic debris derived from erosion of nearby peridotite masses and other more aluminous rocks. If ultramafic debris predominated, lateritic Fe-Ni deposits formed; if other types were more important, Ni-bearing bauxites formed.
3. As above, but with total conversion of the parent material to Fe-Ni laterite or bauxite.

In all of these deposits, oxides predominate over silicates, and pisolitic structures are common. Much of the Ni enrichment occurred after emplacement of the parent material, based on a general increase of Ni and Co with depth (Maksimovic 1978, fig. 4), so that the ores themselves are not allochthonous and should, where the parent ultramafic material is gone, be classed as crypto-autochthonous.

Theories of Origin

Like Al, there is no argument that lateritic Ni deposits form by tropical or subtropical weathering over fairly long periods of time. Unlike Al, Ni deposits require a specific parent rock, Ni-rich ultramafics. Mineralization is enhanced if the parent rock is highly fractured and only moderately serpentinized (Lelong

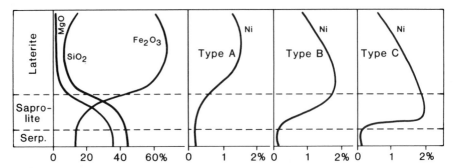

Fig. 4-23. Model for evolution of a Ni laterite with time. Composition (but not thickness) of the soil horizons stays roughly constant, but Ni is gradually transferred lower in the profile because of its moderate solubility in the oxide zone. Thus oxide (A), intermediate (B), and saprolite (C) ores are stages in an evolutionary sequence (after Schellmann 1971, fig. 1).

and others 1976, p. 145); complete serpentinization creates a relatively imperme-able rock mass. The most favorable climate seems to be sub-tropical (e.g., New Caledonia) because in the tropics leaching is too fast to permit formation of thick saprolite horizons that can trap the Ni being leached from the oxide horizons.

Time may also be a factor, as suggested by Schellmann (1971). In his model (Fig. 4-23), nickel laterites undergo a progressive enrichment with time, the grade increasing and the position of maximum nickel concentration moving deeper in the profile because of the lower solubility of Ni in silicates than in oxides. Young (Type A) profiles exhibit only relative enrichment in the oxide zone, while mature (Type C) profiles have a strong absolute enrichment of Ni in the saprolite zone. Evidence supporting this scheme can be found in the deposits at Greenvale in Queensland, Australia (Burger 1979). Here, nickel laterites are found on three terraces. The highest, and presumably oldest, show the most intense mineralization (Fig. 4-24). Furthermore, all of the oxide zone and much of the weathered serpentinite has been removed by erosion, leaving the Ni enrichment deep in the saprolite zone. On the youngest level, a complete vertical sequence is preserved (Burger 1979, fig. 3) and Ni grades greater than 1 percent are found in the base of the oxide and at the top of the saprolite zone.

Summary

Ni laterites form in much the same way as bauxites, but are more constrained by the parent rock, which is invariably an ultramafic or, occasionally, clastic debris eroded from an ultramafic body. Another distinguishing feature is the presence of two zones of mineralization with contrasting mineralogy—the oxide and saprolite zones. It appears that, during the evolution of a deposit, miner-alization moves from the oxide into the saprolite zone.

The principal problems in lateritic nickel deposits involve mineralogy and

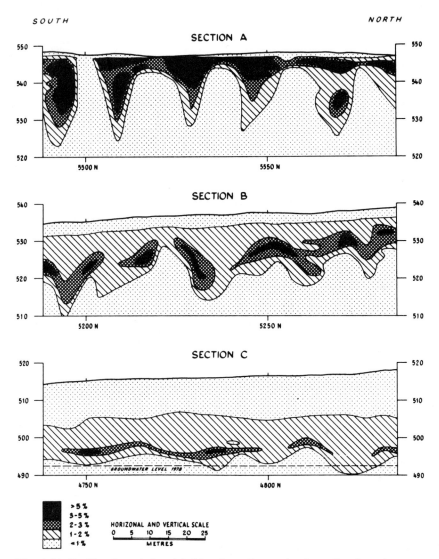

Fig. 4-24. Profiles through the nickel laterite at Greenvale, Australia, show three stages of evolution preserved on different terrace levels. The upper (oldest) terrace has the thickest nickel accumulation, which lies deep in the saprolite zone. The limonite horizon has been completely eroded from this level. The lower levels have a more complete profile with thinner nickel accumulations (Burger, 1979, fig. 4).

petrography. What, in fact, is the distribution, laterally as well as vertically, of the minerals and their textural varieties? Specifically, what controls the distribution of Ni-rich talc? What phase holds the Ni in the oxide zone? Can the thermodynamic relations between Ni-Mg solid solutions and aqueous solutions be determined?

CHAPTER 5.

Manganese

"He is Poseidon's head man and knows every inch of ground all
over the bottom of the sea. If you can snare him and hold him
tight, he will tell you about your voyage, what course you are to
take, and how you are to sail the sea so as to reach your home."

Idothea to Menelaus in *The Odyssey*, Book IV.
From Samuel Butler (trans.), *The Odyssey of Homer*.
(Roslyn, New York: Walter J. Black, Inc.,
for the Classics Club®, 1944, p.47.)
Reprinted with permission of the publisher.

Manganese is one of the more abundant elements in the crust, and enrichments
are accordingly common. Most production, however, comes from a few giant
deposits, particularly Nikopol in the Ukraine, which contains some 75 percent
of the world's reserves (Table 5-1). Also, note another large group of deposits
in the Proterozoic, that occur at the same time as the largest iron ore accumulation.
Geochemically, manganese behavior is controlled by the oxidation state of the
environment. A number of Mn oxidation states are known from laboratory experiments, but only two, Mn^{2+} and Mn^{4+} are common in nature. They often
occur together in the same mineral or in closely intergrown mineral aggregates
giving a continuous series of net oxidation states. The geochemical properties
of manganese are similar to those of iron, so that its segregation into economic
deposits is largely a problem of separating it from iron.

There are a variety of types of manganese deposits. These can be syngenetic
in clastic or carbonate sequences, or in volcanic-sedimentary accumulations, or
they can be supergene deposits in which the original type of accumulation is
obscured by surficial oxidation. A special example of the syngenetic type is the
extensive pavement of ferromaganese nodules that cover the floor in parts of
many deep ocean basins. These are not now commercial, but consitute an important future resource for Cu, Ni, and Co as well as Mn. In this chapter, we
will deal primarily with those ores that form syngenetically in sediments. The
volcanic type is discussed in Chapter 9.

Manganese is twelfth in abundance in the elements of the Earth's crust, just
after phosphorus. As would be expected, based on its similarity to iron, it is
much more abundant in basic than in acid igneous rocks, but its ratio to Fe is

Table 5-1. Distribution of world reserves of manganese.

	10^6 tons of Mn	Percent
Quaternary	70	6.6
Oligocene (Ukranian deposits)	795	74.9
Paleocene (North Urals)	10	0.9
Jurassic-Cretaceous (Morocco)	3	0.3
Carboniferous (China)	15	1.4
Devonian (Kasakhstan)	10	0.9
Silurian (Brazil)	3	0.3
Cambrian (mostly Siberia)	45	4.2
Proterozoic (mostly Brazil, South Africa, China)	90	8.4
Archean (India)	20	1.9

Source: After Varentsov 1964, fig. 13.

virtually constant (Table 5-2). Accordingly, magmatic processes are not important in producing Mn ores because they do not produce a separation from Fe. Sedimentary rocks, on the other hand, show considerable variation in Mn/Fe. Note the high Mn content and high Mn/Fe ratio in pelagic clays. Carbonate rocks have the highest Mn/Fe ratio, although the absolute amount of Mn they contain is not especially high. Many smaller Mn deposits are, in fact, hosted by carbonates. The abundance in shale shows an interesting pattern: organic-rich shales are sharply depleted in Mn, especially relative to iron, in contrast to most other metals, which tend to be strongly enriched in black shales. This behavior can be explained by the higher solubility of Mn than other metals under reducing conditions, as explained in the section on geochemistry.

Table 5-2. Abundance of Mn in common rocks.

	ppm MnO	Mn/ΣFe
Igneous Rocks		
granites	260	0.015
granodiorites	390	0.017
diorites	1390	0.019
gabbros	1390	0.016
peridotites	1050	0.016
Sediments and Sedimentary Rocks		
graywackes	690	0.020
quartz sandstones	170	0.03
shales	600	0.013
black shales	150	0.008
limestones	550	0.12
deep-sea clays ($CaCO_3$-free)	5700	0.095
seawater	0.0013	

Source: After Wedepohl 1980, tables 2, 3.

Table 5-3. Manganese minerals.

Name	Formula	Comments
Tetravalent oxides		
pyrolusite	β-MnO_2	Single chains of $(MnO_6)^{-8}$ octahedra
ramsdellite	MnO_2	Double chains
nsutite	γ-MnO_2	Intergrowths of above
hollandite	$(Ba,K)_{1-2}Mn_8O_{16}\cdot\chi H_2O$	
cryptomelane	$K_{1-2}Mn_8O_{16}\cdot\chi H_2O$	Have open tunnels permitting incorporation of large cations
psilomelane (romanechite)	$(Ba,K,Mn,Co)_3(O,OH)_6Mn_8O_{16}$	
todorokite ("10 Å manganite") (buserite)	$(Na,Ca,K,Ba,Mn)Mn_3O_7\cdot\chi H_2O$	
birnessite ("7 Å manganite")	$(Ca,Na)(Mn^{2+},Mn^{4+})_7O_{14}\cdot\chi H_2O$	Two-layer structure
δ-MnO	$(Mn,Co^{3+})Mn_6O_{13}\cdot\chi H_2O$	Disordered birnessite
lithiophorite	$(Al,Li)(OH)_2\cdot MnO_2$	Two-layer structure
rancieite	$(Ca,Mn)Mn_4O_9\cdot 3H_2O$	Poorly known
Trivalent oxides α-$(Mn,Fe)_2O_3$		
partridgeite	0–10% Fe_2O_3	
situparite	10–30%	
bixbyite	>30%	
Spinel-type oxides		
hausmannite	Mn_3O_4	Structure like magnetite; up to 7 mole % Fe_3O_4
jacobsite		More than 46 mole % Fe_3O_4
vredenburgite	$(Fe,Mn)_3O_4$	Intergrowths of hausmannite and jacobsite with <46% Fe_3O_4
Wustite-type oxides		
manganosite	$Mn_{1-x}O$	NaCl structure, rare
Hydroxides		
pyrochroite	$Mn(OH)_2$	White; brucite structure
manganite	γ-MnOOH	Oxidizes readily to pyrolusite
groutite	α-MnOOH	Uncommon
Silicates		
braunite	Mn_7SiO_{12}	SiO_2 up to 40 weight %
neotocite	$(Mn,Fe)SiO_3\cdot nH_2O$	
Carbonates		
rhodochrosite	$MnCO_3$	Up to 20 mole % Ca
kutnahorite	$CaMn(CO_3)_2$	

Source: After Heubner 1976; Burns and Burns 1977a; Frenzel 1980. For details of X-ray patterns see Burns and Burns 1977b and Frenzel 1980.

Table 5-4. Average microprobe determinations of composition of manganese minerals (ancient deposits).

Groote Eylandt, Australia.[1] Unmetamorphosed deposits in Cretaceous sands and clays. Small amounts of todorokite and nsutite also found.

	MnO_2	Fe_2O_3	K_2O	BaO	Al_2O_3	SiO_2
pyrolusite	98.5	0.7	0.3	0.2		
cryptomelane:						
isotropic	90.2	1.0	3.0	1.5		
anisotropic	90.2	0.3	3.7	0.3		
lithiophorite (Li_2O = 1.26)						
massive	51.9	5.2	0.3	—	22.1	1.9
recrystallized	60.2	0.3	0.1	—	22.0	0.2
psilomelane	85.6	2.7	1.1	5.7	0.7	0.1

Andhra Pradesh, India.[2] High-grade metamorphic deposits in sillimanite-garnet gneiss and quartzite (all Mn recalculated as MnO_2).

	MnO_2	Fe_2O_3	K_2O	BaO	CaO	MgO	Al_2O_3	SiO_2
primary oxides								
bixbyite	41.9	56.6	—	—	—	0.4	1.1	—
braunite	81.4	8.4	—	—	—	—	0.1	10.3
hollandite	72.7	11.9	0.1	10.5	1.2	0.4	2.2	0.9
hausmannite	92.7	6.6	—	—	—	—	0.7	0.1
low-Fe jacobsite	51.3	47.6	—	0.2	—	0.3	0.4	0.1
high-Fe jacobsite	23.2	72.0	—	0.7	—	—	3.9	—
vredenburgite	67.5	28.3	—	0.6	—	0.7	2.8	0.4
Secondary Oxides								
pyrolusite	95.3	1.4	—	—	0.1	—	—	0.8
cryptomelane	91.2	1.4	5.3	0.9	0.3	—	0.4	0.3
psilomelane	79.6	1.8	0.7	13.7	0.5	—	0.6	0.5

Sources: 1. Ostwald 1975.
          2. Sivaprakash 1980.

Mineralogy

The mineralogy of manganese is complex and analytically difficult. Not only are a large number of phases commonly found, but they are difficult to characterize because of poor crystallinity, fine grain size, intimate intergrowths, and a propensity for alteration during sample handling. Oxides are the most common ore minerals. Burns and Burns (1977a, b) have suggested that manganese oxides are made up of subunits consisting of MnO_6^{8-} octahedra. These are then arranged in chains or sheets, much as in silicate structures, to give the observed mineral forms. Note that in many cases the chains are arranged in such a way as to form large "tunnels" like those in zeolites. These spaces may be occupied by large cations such as K^+, Ca^{2+}, or Ba^{2+} (Table 5-3). Some representative microprobe analyses of Mn oxides are given in Table 5-4. Complete descriptions of Mn phases, along with X-ray spacings and photomicrographs of polished sections, can be found in Frenzel (1980).

Many deposits contain Mn carbonates, usually described as rhodochrosite, but analyses show the presence of considerable Ca. Experimental work (Goldsmith and Graf 1957) reveals that there is a solubility gap in the Mn-Ca carbonates between 50 mole percent Mn (the mineral kutnahorite) and 80 mole percent Mn (calcian rhodochrosite), but many analyses of Mn-carbonates from modern sediments (Table 5-5) show Mn contents within this gap, suggesting that they should be metastable with respect to kutnahorite and rhodochrosite. Kutnahorite is not reported from Mn ores but perhaps has not been looked for.

Changes in mineralogy during metamorphism have been described from the Kalahari manganese field of South Africa (Table 5-6). Bixbyite, manganite, hausmannite, and a silica-deficient braunite are characteristic of the metamorphosed zone. Heubner (1976) suggested that the persistence of such minerals indicates the presence of locally high oxygen fugacities during metamorphism; manganosite is the only stable phase under conditions prevailing during the metamorphism of common ferruginous rocks. The Kalahari deposits were later subjected to supergene alteration which resulted in the formation of a distinctive

Table 5-5. Composition of Mn carbonates from recent sediments.

Locality	Mn	Ca	Fe	Mg
		Mole Fraction		
Marine				
Baltic Sea	0.70	0.30	—	—
	0.60	0.32	—	0.08
Loch Fyne, Scotland	0.48	0.45	—	0.07
Freshwater				
Green Bay, Wisconsin	0.73	0.27	—	—
Pinnus Yarvi, USSR	0.51	0.45	0.04	—
	0.54	0.09	0.37	—

Source: After Callendar and Bowser 1976, table 8.

Table 5-6. Manganese minerals from the mines in the Kalahari Manganese Field, South Africa. Formulas are all approximate and the stage of appearance may be different in different localities.

Original sediment
    oolitic calcite
    manganese and iron hydroxide gel
    siderite
    rhodochrosite

Minerals present at diagenetic stage
wad	amorphous
cryptomelane	$K_2Mn_8O_{16}$
braunite I	$\sim 3Mn_2O_3 \cdot Mn_{0.62}Fe_{0.08}Ca_{0.30}SiO_3$
jacobsite	$Fe_2O_3 \cdot MnO_3$
rhodochrosite	$MnCO_3$
calcite	$CaCO_3$

Minerals present at metamorphic stage
  Low grade
braunite I	$\sim 3MnO_2O_3 \cdot Mn_{0.62}Fe_{0.08}Ca_{0.30}SiO_3$
jacobsite	$Fe_2O_3 \cdot MnO_3$
cryptomelane	$K_3Mn_8O_{26}$
pyrolusite	MnO_2

  High grade
hausmannite	Mn_3O_4
jacobsite	$Fe_2O_3 \cdot MnO$
bixbyite	$(Mn,Fe)_2O_3$
braunite II	$\sim 3Mn_2O_3 \cdot Fe_{1.4}Ca_{0.6}Si_{0.5}O_3$
manganite	$MnO(OH)$

Minerals formed during later hydrothermal alteration
acmite	$NaFe^{3+}Si_2O_6$
cymrite	$Ba_2Al_4Si_4O_{15}(OH)_2 \cdot H_2O$
piedmontite	$Ca_2(Al,Fe,Mn)_3(OH)Si_3O_{12}$

Supergene minerals
cryptomelane	$KMn_8O_{16} \cdot nH_2O$
nsutite	γMnO
todorokite	$\sim(Mn_7Mn_{5.2}Mg_{0.4})(Ca_{0.4}Na_{0.2}K_{0.05})O_{12} \cdot nH_2O$
pyrolusite	MnO_2
lithiophorite	$(Al,Li)MnO_2(OH)_2$

Source: de Villiers 1971, p. 56.

assemblage including nsutite, todorokite, and lithiophorite. Local control of oxygen fugacity during metamorphism is also indicated in Sivaprakash's (1980) study of the gondite deposits of Andhra Pradesh, India. From the compositions of the phases (Table 5-4), he calculated a peak temperature of 700°C with a wide range of oxygen fugacities: 10^{-2} to 10^{-12} atmospheres.

Geochemistry

Like other transition elements, Mn is subject to crystal field effects, but only Mn^{4+}, and to a lesser extent the rare Mn^{3+}, have a crystal field stabilization energy. Accordingly, Mn^{4+} in octahedral positions in minerals should be strongly favored over Mn^{2+} in solution (Crerar and others 1980a, p. 296). Those minerals that do contain Mn^{2+}, such as rhodochrosite, tend to be light colored, compared with the dark Mn^{4+} minerals, again because of the splitting of d orbitals of Mn^{4+} in the imposed crystal field, which gives rise to excited states with the same spin multiplicity.

As with iron, manganese geochemistry in sedimentary environments is governed by oxidation and reduction. Eh-pH relations show a relatively large field of stability for dissolved Mn^{2+} (Fig. 5-1). At the pH of seawater (8) or of fresh surface waters (5-7), Mn should be soluble except under strongly oxidizing conditions. Unfortunately, we lack thermodynamic data for many of the most common Mn minerals reported from ores and deep-sea nodules, for instance

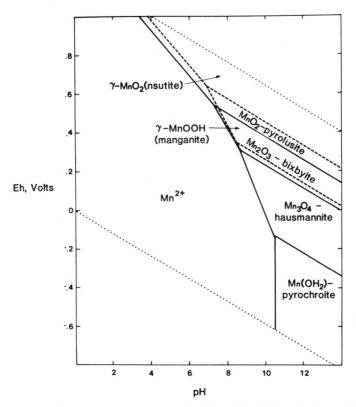

Fig. 5-1. Eh-Ph diagram for Mn oxides, showing Mn to be soluble at low Eh. The dashed lines give the fields for some metastable but commonly encountered phases $(a_{Mn^{2+}} = 10^{-6})$.

todorokite and birnessite, but it is presumed that their behavior is qualitatively similar to that of the simpler oxides.

The addition of sulfur to the system does not change this picture appreciably because of the small stability field for MnS (albandite), but the addition of carbonates creates a large region in which solid Mn species are stable under reducing conditions (Fig. 5-2). Thus, Mn behavior at low Eh is controlled by carbonate minerals, in contrast to Fe, which is controlled more by the sulfides. For fresh water, the pH is normally too low for rhodochrosite precipitation, but for seawater, a slight increase in the amount of CO_3^{2-} should lead to rhodochrosite formation if there is a supply of Mn. This mineral has, in fact, been reported from low Eh marine environments with a variety of water depths and sedimentation rates (Calvert and Price 1970, Suess 1979, Pedersen and Price 1982).

As first pointed out by Krauskopf (1957), Eh-pH diagrams suggest a mechanism for separating Mn from Fe. Figure 5-3 shows that soluble Mn has a considerably larger stability field than soluble Fe under moderately reducing conditions. Because many sediments become reducing a few centimeters below the sediment-water interface, there should be a mobilization of Mn^{2+} into the pore water,

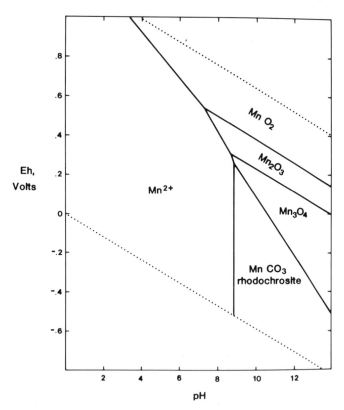

Fig. 5-2. Eh-pH diagram for Mn oxides and carbonates. The sulfide (albandite), not represented here, occupies a smaller field than that of iron sulfides, and is largely eliminated by the field of rhodochrosite. ($a_{Mn^{2+}} = 10^{-6}$; $a_{CO_3^{2-}} = 10^{-3}$).

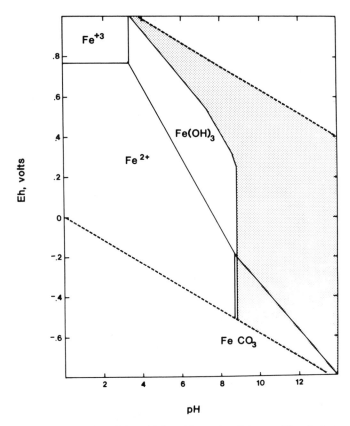

Fig. 5-3. Relative solubility of Fe and Mn oxides and carbonates. The shaded area shows the fields of stability of Mn minerals from Fig. 5-2.

while iron remains fixed as an oxide or hydroxide. Under conditions of low Eh and high sulfur content, such as in reducing marine sediments, the iron would be fixed as the sulfide. This dissolved Mn^{2+} diffuses upward, either to be precipitated at the interface by the oxygen in the bottom water (Fig. 5-4), or else dispersed into the overlying water and carried into deeper parts of the basin (Fig. 5-5), depending on the oxygen content of the water (Yeats and others 1979). This latter process is one of the mechanisms by which Mn becomes enriched in pelagic clays (Table 5-2). Another is alteration of basalts at mid-ocean ridges. Hydrothermal refluxing of seawater through the oceanic crust at the ridges releases large amounts of Fe and Mn to the overlying seawater; most of the Fe precipitates near the vents, sometimes with Cu sulfides, but the Mn shows a considerably greater dispersion (Lupton and others 1980), which can be used as a geochemical tracer for certain types of ore deposits (Cronan 1980, p. 278-282). This aspect of Mn behavior is discussed more fully in Chapter 9.

Such diagrams can also be used to explain supergene enrichment of ferromanganese ores. The greater mobility of Mn will lead to a greater tendency for it to move downwards in soil profiles, leaving behind an Fe-rich crust. If the

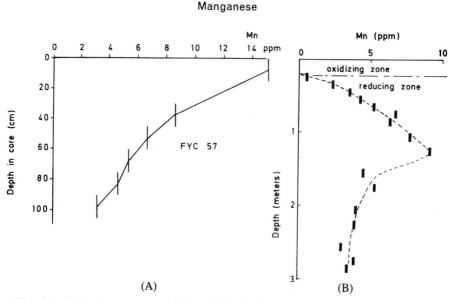

Fig. 5-4. Typical pore-water profiles of dissolved Mn. Left (Loch Fyne, Scotland):
formation of Mn carbonate depresses Mn concentration leading to downward
diffusion. Right (Arctic Basin): more common form shows mobilization of Mn
in anoxic zone, most of which diffuses upwards (Wedepohl 1969, fig. 25-I-
2).

Mn is not lost from the soil altogether, its downward movement can produce
significant enrichments (Sivaprakash 1980, p. 1102; Roy 1981, p. 124-132).
Excluding the Russian deposits, the great majority of mineable Mn ores have
undergone supergene enrichment (Lelong and others 1976, p. 126).

The variety of Mn phases found in sediments and the common presence of
metastable ones suggests that kinetic factors may be important in Mn geochem-

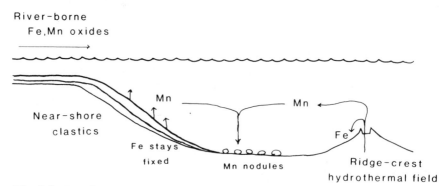

Fig. 5-5. Transfer of Mn from shallow-water clastic sediments and from mid-ocean ridge
basalts to deep water. Not to scale.

istry. Unlike most other metals, some work has been done on the geochemical kinetics of Mn (Hem 1963, 1981; Stumm and Morgan 1970, p. 525-555; Glasby 1974, p. 15-18). For most manganese oxides, precipitation involves oxidation of the Mn^{2+} in solution to Mn^{4+}. Although thermodynamically favored, as explained above, this transition is strongly inhibited kinetically. Whereas oxidation of Fe^{2+} can be modelled by the expression

$$d(Fe^{2+})/dt = k(Fe^{2+})(O_2)(OH^-)^2,$$

Mn^{2+} oxidation follows an autocatalytic relationship:

$$d(Mn^{2+})/dt = k_1(Mn^{2+}) + k_2(Mn^{2+})(MnO_2)(O_2)(OH^-)^2.$$

This expression tells us that the precipitation of Mn oxides will be favored by the presence of a surface of manganese oxide, or as it turns out, of iron oxide. These oxide surfaces have an ability to sorb appreciable quantities of ions from solution, particularly favoring the cations of transition metals. Oxidation of Mn^{2+} involves, first, a sorption of the ion onto the surface, followed by the oxidation step; thus, the importance of a pre-existing surface. Mn and Fe oxidation also differ in their pH dependence: Fe oxide forms at an appreciable rate at pH's above 6, while the equivalent rate for Mn oxide is not reached until pH = 8.5 (Stumm and Morgan 1970, fig. 10-7d). Thus, in seawater (pH = 8), there will be a tendency for Fe to precipitate, but for Mn to remain in solution, even when thermodynamic considerations suggest that both should precipitate, so that the catalytic effect of oxide surfaces is especially important in seawater. In fact, most manganese nodules appear to have nucleated around some grain, such as a shark's tooth, and these nuclei commonly have a rim of iron oxide or hydroxide that precedes the deposition of the Mn oxide layers (Burns and Brown, 1972). The constants in these equations were determined in batch experiments and are diffucult to apply to natural systems with flowing water, but Hem (1981) has calculated that, at pH 8.5, it would take nearly a million years to form an oxide layer 0.1 mm thick. Note that these kinetic constraints do not apply to Mn carbonates, which form directly from Mn^{2+}. Balzer (1982, fig. 3) has shown experimentally that Mn^{2+} concentrations in bottom waters of the Baltic rise above the level predicted by equilibrium with oxides as the underlying sediment becomes anoxic, and stabilize at about the level predicted for $MnCO_3$.

The high capacity of manganese oxides for the sorption of cations leads to an enrichment in a number of economically valuable transition metals, particularly Cu, Ni, and Co. But the exact mechanism of incorporation remains controversial. For example, Burns and Burns (1977a, p. 297-298) maintained that these elements substitute within the lattice of the manganese minerals, while Glasby (1974, p. 23-34) has argued that such cannot be the case and instead the metals are loosely held on exchange sites. Enrichment is much greater in the pelagic nodules than in shallow-water or lacustrine nodules, perhaps because of slower deposition (Table 5-7). Trace element enrichments similar to those in fresh water nodules are found in commercial manganese deposits, but note the much lower Fe contents. Typically, the Fe concentration averages 1 to 2 percent, rarely exceeding

Table 5-7. Concentration of some trace elements in accumulations of ferro-manganese oxides.

Location	Percent		ppm				
	Mn	Fe	Co	Ni	Cu	Zn	Pb
Pelagic Nodules	20	20	5000	5000	500	500	1000
Near-Shore Nodules	30	10	50	50	50	50	50
Fresh-Water Nodules	20	15	200	200	600	1600	500
Manganese Ores:							
Nikopol, USSR	45	1.0	30	200	70	—	—
Ouarzazate, Morocco	40	0.4	200	60	1300	900	—
Singay, Philippines	—	2.5	500	50	10	250	—
Lucifer, Mexico	50	0.7	350	20	2700	1200	6400
Eregli, Turkey	—	1.2	25	50	90	—	—

Source: From Hewett 1966, table 7; Glasby 1974, table 1; Callender and Bowser 1976, table 7.

10 percent. Thus, even the shallow-water accumulations of manganese nodules are not direct analogues for the formation of the ore deposits.

Understanding the influence of microorganisms is another problem in the geochemistry of manganese. Bacteria catalyze the oxidation of Fe^{2+} (Stumm and Morgan 1970, p. 542), particularly at low pH's where the abiotic reaction is slow, and certain bacterial species are associated with Mn oxides, but whether or not they influence reactions of Mn has been debated (Glasby, 1974, p. 19-20). A number of recent papers on the bacteriology of Mn nodules are presented in Krumbein (1978). None of the work so far demonstrates an essential role for bacteria in the development of nodules, but some evidence suggests that they may be important during nucleation. Furthermore, the slow precipitation rates for Mn oxides cited above are greatly exceeded in some modern environments (e.g., Emerson and others 1979, Crerar and others 1980b) suggesting bacterial catalysis.

Petrography

Although Mn ores usually exhibit a variety of concretionary or oolitic textures, there are only a few detailed accounts of their petrography. One well-described example is the Proterozoic ores of Nsuta, in West Africa. The sequence, deposited about 1800 million years ago, contains siderite-greenalite banded iron-formation as well as the manganese beds (Leclerc and Weber 1980). The ores are supergene enrichments of manganiferous black shales with primary Mn contents of 10 to 25 percent. Mineralogically, the shales consist of illitic clays, organic matter, carbonates, and pyrite. The Mn is found in the carbonates, which occur in three textural types: disseminated small crystals, irregular pisolitic aggregates, and isolated clear rhombohedra that usually have corroded edges. The first two types are Ca-Mn carbonates with Mn ranging from 40 to 70 percent, while the last is an Mn-free dolomite. The pyrite grains are surrounded by yet another compositional variety, an Fe-Mn-Mg carbonate. Ni, Co, and Cu are also present in small amounts, and, although for the sequence as a whole they correlate positively with Mn and negatively with Fe, microprobe images show them to be exclusively associated with the pyrite in the Mn-rich beds, which Leclerc and Weber (1980, p. 106) interpreted as indicating a substantial rearrangement of these elements during diagenesis. Possibly the Mn was first deposited at the sediment surface as an Mn^{4+} oxide incorporating Co, Ni, and Cu. Then, during early diagenesis, conditions became more reducing, and Mn carbonates grew at the expense of the oxides (Fig. 5-2), releasing the transition metals which were incorporated in the pyrite that was growing at the same time. After SO_4^{2-} was exhausted, pyrite formation ceased and Fe-Mn carbonates were precipitated around the pyrite. If this paragenetic scheme is correct, one would expect isotopically light carbon, derived from the oxidation of organic matter, in the Ca-Mn carbonates, but isotopically heavy carbon, derived from methane fermentation, in the later Fe-Mn carbonates (Irwin and others 1977).

The ores were then subjected to secondary enrichment during lateritic weathering in the Tertiary (Roy 1976, p. 449-450; see also Lelong and others 1976 for a description of the environment of weathering). In the lower parts of the weathering profile, below the zone of oxidation of pyrite, replacement begins as manifest by the conversion of all of the carbonates to rhodochrosite (Leclerc and Weber 1980, p. 97). In higher levels, the $MnCO_3$ replaces other phases until a layer of completely replaced shale, a few centimeters thick, forms. This layer is succeeded by a few millimeters of manganite, then 2 to 5 cm of massive pyrolusite (Bricker 1965, fig. 37 and 38). Above this transition zone, the ore is entirely oxides, which display a variety of textural features, divided into three types: replacement, cavity-filling, and residual (Sorem and Cameron 1960). Replacements of micas, garnets, and pre-existing Mn-rich ooliths are the most common features. In replacements, nsutite (γ-MnO_2) predominates, accompanied by goethite, cryptomelane, and occasionally lithiophorite. Cavity-fillings exhibit a variety of laminated, colloform textures. Nsutite again predominates, commonly interlaminated with cryptomelane or pyrolusite. Lithiophiorite occurs as a filling in veins, and occasionally in other cavities. Residual ores were not described, but seem to consist of leached and brecciated parent rock. The paragenetic sequence, although seldom found complete, comprises

1. replacement of rock fragments by nsutite,

2. filling of interstitial spaces by

    a. deposition of alternating layers of cryptomelane and nsutite or

    b. deposition of pyrolusite, and

3. deposition of lithiophorite in fractures or remaining pore space.

The sequence is commonly interrupted and the progression of cavity fillings is variable, except that lithiophorite is never folowed by another Mn mineral. The ore is overlain by a 5 to 6 m unproductive zone of Mn-rich pisoliths in a matrix of yellowish goethite and gibbsite. The pisoliths contain alternating concentric laminae of lithiophorite, goethite, and gibbsite (Leclerc and Weber 1980, p. 105).

Volcanic-sedimentary deposits with less extensive supergene alteration are found in the Olympic Peninsula of Washington (Sorem and Gunn 1967). Here, the supergene suite is dominated by Ca-bearing oxides such as rancieite and birnessite. Nsutite is also common, as is the silicate neotocite. The differences in the supergene assemblage from those of west Africa may reflect less intense leaching or a parent material with higher Ca and Si in the Olympic deposits.

At Groote Eylandt, off the north coast of Australia, deposition under shallow marine conditions has produced a dominantly pisolitic ore (Ostwald 1975, 1980; Slee 1980). Mineralogically, cryptomelane dominates with somewhat smaller amounts of pyrolusite and manganite. Psilomelane and braunite are present but rare, and lithiophorite occurs as a weathering product. Petrographically, the ores have been divided into three types: massive, consisting of fine-grained crypto-melane interbedded with detrital clays that show evidence of replacement by Mn oxides; concretionary, with globular bodies a few centimeters to a few meters in diameter, set in a loose matrix of sandy clay; and pisolitic, the most common variety, comprising roughly spherical, concentrically laminated particles 2 to 15 mm in diameter, set either in a matrix of sandy clay or cemented by Mn oxides. The laminations in the pisoliths are defined by alternating layers of very fine-grained acicular cryptomelane and somewhat coarser pyrolusite. Some of the pisoliths have nuclei of abraded pieces of former pisoliths, indicating that they are truly accretionary, formed in an agitated environment, rather than being microconcretions formed in place in the sediment or in a soil, like those in the Gabon deposits. The absence of Mn carbonate minerals is notable, but Slee (1980) has suggested that these ores are derived from pre-existing carbonates. Unconformably underlying the ore beds is a formerly much more extensive bed of manganiferous marl, which, in his model, was eroded to provide abundant dissolved Mn, which was carried by rivers and reprecipitated at the shoreline. Figure 5-1 shows that, under strongly oxidizing conditions, a rise in pH from 6 to 8, typical of the fresh water to marine transition, should precipitate MnO_2. The preponderance of cryptomelane is a distinctive feature of these ores, and suggests that somehow K^+ must have been abundant at the time of deposition.

For modern-day Mn nodules, several excellent petrographic studies are available; for example Sorem and Foster (1972) and Burns and Brown (1972). Sorem and Foster illustrate the styles of lamination exhibited by nodules. They describe and illustrate massive, mottled, columnar, compact, and finely laminated varieties. Using these, they were able to show "stratigraphic" correlations among three nodules from the same locality. Burns and Brown used optical, X-ray, and microprobe techniques to document the presence of an initial iron oxide/hydroxide coat around the nucleus of Mn nodules.

More recently, Sorem and Fewkes (1979) have published an extensive atlas of textural features of Mn nodules. Techniques for preparing samples for microscopy and X-ray diffraction are described in detail, making this a valuable resource for anyone studying Mn deposits. They report on both external and internal textures. Externally, the nodules are classed as either smooth or granular. X-ray investigation shows that the smooth textures are associated with amorphous, Fe-rich material, and the granular textures with micro-crystalline Mn oxides. Both textural types commonly occur on the same nodule, with the smooth on top. Apparently, the controlling factor is contact with the sediment. Small nodules, which grow mostly within the sediment, and the bases of the larger nodules both show granular, crystalline texture, while the tops of the larger nodules, growing entirely within the water column, usually show the smooth, amorphous texture. Internal structures are also separable into amorphous and crystalline types using the ore microscope—the crystalline material shows a distinct anisotropism in polarized light. X-ray studies reveal that the isotropic material is, in fact, amorphous. The distinctive "columnar" structure, which has an appearance much like that of stromatolites, is characteristic of Fe-rich amorphous oxides, as is massive, high reflectivity "compact" structure. Crystalline Mn oxides, usually intimate intergrowths of todorokite and birnessite, form micro-dendrites that appear as "mottled" texture in plan view. In section, this dendritic material often passes up into large, massive "pod" structures. The crystalline portions of the nodules contain inclusions of fine clastic debris and microfossils which are rare in the amorphous portions. Sorem and Fewkes speculate that crystalline oxides form only where the nodule contacts the surrounding sediment. In some nodules these structures are disrupted by later boring, which also results in "mottled" textures (Furbish and Schrader 1977).

Vertical Sequence

Manganese enrichments are found in a variety of geologic situations (Table 5-8) that presumably reflect different modes of development. I have selected three districts to illustrate the kinds of vertical sequence found: clastic-hosted deposits of the Ukraine, the carbonate-hosted ores of Morocco, and the iron formation-related deposits of Gabon. Ores of the volcanic-sedimentary type are discussed in Chapter 9; metamorphosed equivalents of these types have been described by Roy (1981, p. 323-345).

Table 5-8. Types of Mn deposits, classified by associated lithologies.

Type	Example	Reference
Deep-sea	Pacific Mn-nodules	Glasby 1977
Clastic-hosted	Nikopol	Varentsov and Rahkmanov 1977
Carbonate-hosted	Morocco	Bouladon and Jouravsky 1952
Volcanic-hosted	Jalisco, Mexico	Zantop 1978
Metamorphosed	Kalahari	deVilliers 1971

Nikopol

Between 75 and 80 percent of the world's present reserves of Mn are contained in the deposits at Nikopol and Chiatura in the USSR. This giant accumulation was formed over a relatively brief period in the Early Oligocene. Most descriptions are in Russian, but several reviews have been translated into English (Varentsov 1964, Strakhov and others 1970, Varentsov and Rakhmanov 1977, 1980). In this section, only Nikopol is discussed; the pattern at Chiatura is similar. The ores are found in a single horizon, 2 to 3 m thick in a band about 25 km wide and over 150 km long (Fig. 5-6). The ore thickens where it fills depressions in the underlying basement, and is cut out by later erosion in places. To the north, the ore bed wedges out; to the south, it disappears because of an increase in clay interbeds, and a gradual dilution of the ore by clastics. The vertical sequence (Fig. 5-7) shows that the ore bed formed during transgression onto the crystalline basement. Lying between the ore and the basement, in most places, is a set of thin, glauconitic sands containing shell-banks with numerous *Glycymeris* and *Cyprina*, indicating deposition on beaches (Gryaznov and Barg 1975). Over-stepping these beds is the manganese horizon, which changes from oxides in the north to carbonates in the south, corresponding to increasing depth of water (Fig. 5-7). The top of the ore horizon is marked by iron hydroxides in the oxide ores, but by glauconite in the carbonate ores. The overlying sediments are predominantly fine-grained, greenish-gray, montmorillonitic clays. Their thickness varies from 0 to 25 m depending mostly on the extent of erosion in the Late Oligocene.

The ore horizon itself has been divided into three facies (Varentsov and Rakhmanov 1977): oxide, mixed oxide-carbonate, and carbonate (Table 5-9). The carbonate facies occupies the most distal position, and occurs in two textural varieties: concretionary-nodular ores with nodules 1 to 25 cm in diameter in a clay-silt matrix, and coarse-lumpy ores that have a highly porous texture. Both

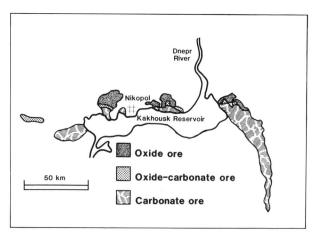

Fig. 5-6. Orebodies of the Nikopol area, showing distribution of facies (after Varentsov and Rakhmanov 1977, fig. 48).

Fig 5-7.
Idealized vertical profile through Mn
horizon at Nikopol (after Varentsov
and Rakhmanov 1977, fig. 49).

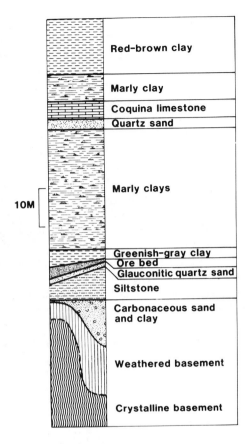

Table 5-9. Composition of the three ore types found in the Nikopol deposits.

Percent	Carbonate	Matrix	Mixed Nodules	Oxide
Mn	23.70	18.3	44.4	51.2
Fe	1.31	1.53	1.07	0.57
P	0.09	0.31	0.25	0.02
CO_2	22.60	25.60	3.73	0.60
C_{org}	0.28	0.13	0.06	0.12
ppm				
Cu	14	16	37	41
Ni	64	67	334	328
Co	10	9	16	34
V	28	37	118	68
Cr	10	8	16	10

Source: After Varentsov and Rakhmanov 1977, p. 124–132.

rhodochrosite and manganoan calcite are reported, always poorly crystalline and very fine-grained. Besides the carbonate minerals, the nodular variety contains sponge spicules, diatoms, and fishbones, while the lumpy variety is relatively free of such inclusions.

Towards the paleoshoreline, this ore type passes into a mixed oxide-carbonate facies. Here, the ore consists of irregular spherical masses, from 0.1 to 1 cm, of Mn oxide in a matrix of Mn carbonates. These oxide segregations make up about 25 percent of the ore. Corrosion and replacement of the oxides by the carbonate matrix suggest that the oxides formed early in the development of the beds, before the carbonates. A similar paragenesis is seen in the $MnCO_3$-cemented oxide nodules of Loch Fyne, Scotland (Calvert and Price 1970, p. 220). The carbonate groundmass in this facies has a composition close to that of the carbonate facies itself (Table 5-9).

In the areas closest to the paleo-shoreline, the ore consists almost entirely of manganese oxides. It has been much debated to what extent this is a primary oxide facies, to what extent an expression of supergene alteration of carbonate ore. Varentsov and Rakhmanov (1977, p. 135) believed that only the oxide-carbonate and carbonate ores are primary, and that the pure oxide facies developed through secondary oxidation of the other two types. Texturally, the oxides in this facies consist of earthy masses or concretions of up to 25 cm across, sometimes with relics of carbonate or oxide-carbonate structures. There is well-developed concentric layering, but with much deformation, and smaller concretions are often coalesced into larger, compound aggregates.

Morocco

Moroccan deposits. Bedded Mn deposits in carbonate host rocks are widespread, although reserves are subordinate to those found in clastic sequences like Nikopol or to those associated with banded iron-formation like Urucum, in Mato Grosso, Brazil (Varentsov 1964, p. 5). In addition to Morocco, deposits of this type are found in Lower Cambrian carbonates of the Southern Appalachians (e.g., Rodgers 1945, Espenshade 1954) and in Precambrian rocks of India (Roy 1981, p. 355-361). In Hungary, similar deposits are found in radiolarian marls in a limestone sequence (Cseh-Németh and others 1980, fig. 5).

The Moroccan ores have been described by Bouladon and Jouravsky (1952, 1956) and Vincienne (1956); a summary of their work can be found in Varentsov (1964, p. 24-35). Mn ores of two types are important in Morocco: vein and volcanic-sedimentary deposits of Precambrian and Early Paleozoic age, and slightly lower-grade syngenetic deposits of Mesozoic age, mostly Liassic and Cenomanian-Turonian. As of 1955, the Cretaceous deposits at Imini had produced over half the total Mn for the country (Bouladon and Jouravsky 1956, p. 219). The presence of older, higher grade deposits, and the repetition of the syngenetic mineralization at several times in the Mesozoic, suggests that the syngenetic deposits are second-cycle, derived from weathering of the older deposits (Vincienne 1956). Some of the ore minerals may even be detrital. Roy (1976, p. 428) has argued that the presence of braunite, hausmannite and jacobsite in some localities indicates redeposition of metamorphic Mn minerals. However, most of the enrichment must have been via solution, otherwise the ore would have been diluted by clastics.

In vertical section (Fig. 5-8), the ore lies between clastics and carbonates, apparently reflecting an intermediate position in a proximal-distal sequence. The clastics are a coarse, near-shore to continental red-bed sequence; the carbonates are mostly fine-grained dolomites, with some thin cherts and remains of the oyster *Exogyra africana*, indicating shallow, agitated conditions. From the red-bed and gypsum association it can also be inferred that the environment of deposition was near the shoreline in an arid climate. Mineralogically, the ores consist of pyrolusite with psilomelane and coronadite, but little is known about the zoning of these minerals or their textural relationships. Thus, although we can surmise the origin of the Mn, and assume that it was carried in solution by

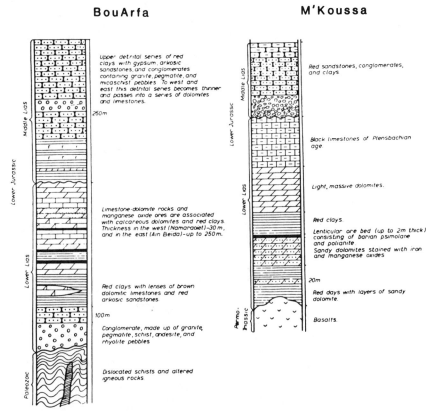

Fig. 5-8. Vertical profiles of Mesozoic Mn deposits of Morocco, showing position of Mn enrichment between clastic and carbonate facies (Varentsov 1964, figs. 8 and 9).

rivers into the basin of deposition, it is not clear what process was responsible for precipitating it at the limit of clastic deposition. That is, if the precipitation were caused simply by the mixing of river water with seawater, the Mn should have been mixed with the near-shore clastics to form much lower-grade deposits.

It is also uncertain what the contribution of supergene enrichment to ore formation has been. Beaudoin and others (1976) have suggested that karstification of the dolomite produced enrichment of a sub-ore grade syngenetic deposit, but Pouit (1976), while accepting the presence of abundant karst features, argued that they were incidental to ore genesis. He further observed that the alignment of the Imini deposit with the paleoshoreline (Fig. 5-9) suggests syngenetic control of mineralization, as does the distribution of associated elements: Pb is high, up to 2 percent, along the proximal (southern) side of the trend, while Fe and P are rich on the distal (northern) side. Just as we have no mechanism for explaining the deposition of the Mn, the origin of this zoning is unknown.

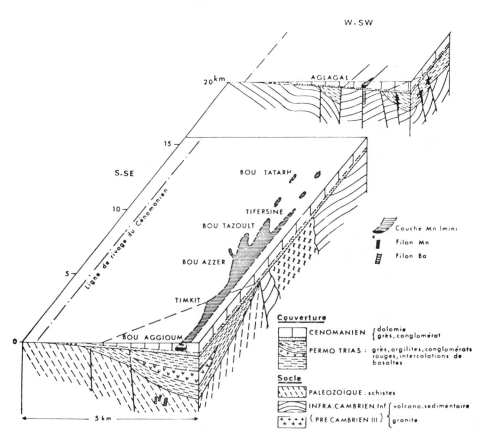

Fig. 5-9. Bedded (couche) ores at Imini are elongated parallel to the paleoshoreline (ligne de rivage), between sand (gres) and dolomite (dolomie) in vertical section. The sequence unconformably overlies older rocks with vein (filon) deposits (after Pouit 1976, fig. 1).

Deposits associated with iron-formations

Mn accumulations are common in the Proterozoic, as might be expected from the volume of iron deposited in that time period. In most cases, however, economic concentrations of Fe and Mn are not found together, or even in the same district, even though subeconomic enrichments normally are associated with ores of the other metal. Metamorphism commonly obscures the depositional features of such Mn deposits, but those of the Francevillian basin of Gabon in west Africa are unaffected by metamorphism, although strongly altered by weathering at the surface. We will use the Gabon deposit as an example; a description of the vertical sequence in the Kalahari Manganese Field of South Africa, which differs from the Gabon deposits in having more limestone, can be found in Beukes (1973, fig. 28). The Gabon ores are contained in a roughly 2000 m sequence of clastics (Leclerc and Weber 1980, fig. 3; Bonhomme and others 1982). These show a general fining-upward trend through about 1000 m from thick conglomerate units at the base to shales, often carbonaceous, which are succeeded by the manganiferous beds. Renewed deposition of coarse clastics followed, again fining upward, but becoming tuffaceous. The Mn beds are not repeated. The beds themselves, from 150 to 200 m thick, contain black shales, iron-formation, and, in some areas, dolomites. The general sequence upward is black shale $\rightarrow$ iron-formation $\rightarrow$ manganiferous black shale. The iron-formation, approximately 10 meters thick, is made up of alternating bands of chert and either pyrite, greenalite, or siderite. Siderite predominates toward the top. This banded iron-formation differs from others, described in Chapter 2, in having high concentrations of phosphate, between 1 and 2 percent (Leclerc and Weber 1980, fig. 5). Above the iron-formation is 100 to 150 m of black, pyritic shale with Mn carbonates. As described in the petrography section, weathering of these carbonates has produced the ores.

Manganese in Modern Sediments

Numerous examples are known of Mn enrichment in modern sediments that can be examined for analogues of the processes involved in the formation of ore in ancient rocks. It is convenient to treat these in two groups, oxides, which form the deep-sea nodules, and carbonates, which are not as extensively developed in the modern, but are of interest because of the abundance of carbonate or carbonate-derived ores in the ancient.

Mn nodules in modern sediments

The world's largest deposit of Mn is found in modern deep-sea sediments. Pelagic ferro-manganese nodules cover the ocean floor over large areas, particularly in the central Pacific. The most economically attractive area, lying between the Clarion and Clipperton fracture zones, has been estimated to contain recoverable reserves of 2×10^9 metric tons averaging 25 percent Mn, 1.3 percent Ni, 1.0 percent Cu, 0.22 percent Co, and 0.05 percent Mo (McKelvey and others 1979). Although Ni and Cu would presumably be the primary target for any future exploitation of this area, it contains 500×10^6 tons of Mn compared with 800×10^6 tons of reserves in the Oligocene deposits of the USSR. The total sea floor must, accordingly, contain several times as much Mn as all the terrestrial deposits.

The mineralogy, geology and geochemistry of Mn nodules has been extensively reviewed (Horn 1972, Cronan 1974, Glasby and Read 1976, Callendar and Bowser 1976, Glasby 1977, Bischoff and Piper 1979, Cronan 1980, Heath 1981). Here, I will mention only a few aspects, concentrating on those that provide comparisons and contrasts with the deposits on land. The most striking feature of the nodule deposits, at least to a geologist, is their lack of a vertical sequence. The nodules usually occur at the sediment-water interface, or lying on top of the sediment. When found within the sediment at greater depths, as in some of the JOIDES cores, they can often be related to unconformities (Glasby and Read 1976, p. 304-305). Their presence at the surface is hard to explain because growth rates obtained from radiometric dating are slower than sedimentation rates of the underlying sediments (Ku 1977, fig. 8-3). Some workers have argued that the radiometric techniques are not applicable to nodules because of frequent hiatuses in growth or because of migration of radionuclides; others have invoked reworking of the sediment by organisms or bottom currents. Although this problem has been known for some years, there is still no agreement on how to resolve it.

The nodules do show interesting areal variations, particularly in minor-element chemistry. Price and Calvert (1970, Calvert and Price 1977) have summarized these data and advanced a diagenetic explanation. In relatively near-shore areas, rates of sedimentation are higher than in central parts of the ocean, resulting in the burial of more organic matter (Berner 1978) and, thus, in lower Eh values within the sediment. As we have already seen, this condition favors the mobilization of Mn, but, often, not of Fe. The soluble Mn then diffuses upward, to be precipitated at the sediment-water interface. Because most of the Fe is left behind, nodules receiving most of their Mn from this diagenetic source should have higher Mn/Fe ratios than the underlying sediment. They also have todorokite as the predominant Mn mineral and lower total amounts of minor metals, but with the bulk of these elements associated with Mn. There is a continuous variation from this type to nodules from central areas, which have an Mn/Fe ratio close to that of the underlying sediment, suggesting that most of their Mn is derived directly from the overlying seawater. These nodules also contain primarily δ-MnO_2 (disordered birnessite), are relatively rich in minor elements, and have more of these minor elements associated with Fe. Piper and Williamson (1977) related high Mn/Fe ratios to high productivity in the overlying surface water, consistent with the idea of mobilization of Mn from organic-rich sediments. Little is known about lateral variations in the composition of fresh-water deposits. In Green Bay, off Lake Michigan, the Mn/Fe ratio appears to increase away from points of stream input because of the more rapid deposition of the iron (Callendar and Bowser 1976, p. 354-359).

Mn carbonates in modern sediments

Because of the preponderance of Mn carbonates as ore minerals or as protores for deposits in ancient rocks, consideration needs to be given to modern-day accumulations of carbonates as well as oxides. Such occurrences have been described from shallow, near-shore sediments of Scotland (Calvert and Price 1970) and the Baltic (Suess 1979) and from deep-water sediments of the Panama Basin (Pedersen and Price 1982). The pore waters of the sediments in the Panama Basin reach Mn^{2+} concentrations as high as 160 micromolar, and the dry sediment reaches as high as 3 weight percent Mn. The high Mn percents are associated with Mn oxides at the sediment surface, but, in one location, a zone of Mn carbonate was also found at a depth of about 150 cm in the sediment. Whitish crusts at this depth proved to be coalesced microspheres of Mn carbonate, about 100 micrometers in diameter, with compositions close to that expected for kutnahorite (Table 5-10). No other diagenetic phases were found.

The Baltic sediments contain a complex diagenetic assemblage that includes siderite, MnS, and iron phosphates, in addition to the Mn carbonates. Mn^{2+} concentrations in the pore waters reach 80 micromolar. Again, the Mn carbonate occurs as microspheres, in this case 5 to 25 micrometers across (Suess 1979, fig. 3), imbedded in a matrix of amorphous silica. The MnS phase has an unusual hexagonal form, as seen with the scanning electron microscope. It is also peculiar that Mn should form sulfides while Fe goes into carbonates and phosphates.

Neither of these two occurrences is particularly reminiscent of ancient deposits, but the Loch Fyne sediments resemble clastic-hosted ores like Nikopol. Manganese occurs both in nodular masses of Mn oxide and as concretions of $MnCO_3$, 1 to 8 cm in diameter. The oxide-rich nodules, in turn, are commonly cemented and replaced by $MnCO_3$ (Calvert and Price 1970, fig. 4 and 5). Over an area of about 10 km^2, Mn concentrations in the surface sediment exceed 5 percent, sometimes reaching 10 percent, although the thickness of this surficial layer is only about 20 cm. Other constituents in the sediment are mostly detrital quartz and clay and shell material. Unlike Nikopol, there do not appear to be distinct oxide and carbonate facies, and the depth of the water, from 180 to 200 m, is greater than seems likely for the Ukraine deposits.

Carbon isotopes (Table 5-10) suggest a variable contribution of organic carbon to formation of the Mn carbonates. The carbon in the Panama Basin carbonates is likely derived entirely from pore water HCO_3^- or dissolution of shell material. The Loch Fyne sediments, from a different environment but with a similar Mn/Ca ratio in the carbonates, have somewhat lighter carbon, indicating some contribution from decaying organic matter. Because well-preserved aragonitic shell material is common in the nodules (Calvert and Price 1970, fig. 4), the balance of the carbon must have come from seawater HCO_3^-. The much more negative values from the Baltic imply that more than half of the carbon is organic-

Table 5-10. Carbon isotopic composition of manganese carbonates in modern marine sediments.

| Locality | Mole Percent | | | δ¹³C, PDB permil |
	Mn	Ca	Mg	
Panama Basin	48	47	5	+ 2.6
Loch Fyne	48	45	7	− 5.8
Baltic Sea	85	10	5	− 13

Source: Pederson and Price 1982, table 3.

derived. Apparently, there is no necessary connection between water depth, type of sediment, or organic activity and the precipitation of Mn carbonates. Ancient deposits need to be investigated to see if they have a similar diversity in isotope geochemistry.

Summary

It is difficult to generalize about the origins of Mn deposits; each type seems to have formed in its own way. The formation of volcanic-sedimentary ores, as described in chapter 9, results from hydrothermal circulation of seawater through hot volcanic rocks. The deep-ocean nodules probably derive some of their Mn from this source after long-distance transport through seawater, but much also comes from mobilization of Mn from clastic sediments with low-Eh porewaters. Deposition of the Mn is favored by the presence of appropriate substrates for nucleation and by extremely slow rates of sedimentation.

Much less is known about syngenetic deposits on land. The most challenging deposit to explain is Nikopol-Chiatura, which, because of its unique size, must have formed as a result of a special set of circumstances. The formation of the ore itself is relatively well understood: the manganese was first deposited directly from seawater in the form of amorphous Mn hydroxides. This material was then converted, in part, into Mn carbonates during early diagenesis by reaction with the abundant organic matter deposited with it (Varentsov and Rakhmanov 1977, p. 134-137). Note that, if the carbonate in rhodochrosite has this source, its carbon isotopes should be very light, by analogy with other diagenetic carbonates (Hudson 1978), although it is possible for rhodochrosite to replace primary calcite (Hager 1980), in which case it would inherit heavier carbon. During subsequent uplift and erosion, the supergene oxides formed.

The source of so much manganese remains puzzling, however. Russian workers believe that Mn was leached from soils developed on the metamorphic basement, leaving iron behind as a weathering crust. This Mn was then carried by streams into the basin where it was precipitated at the shoreline. Although similar to the mechanism described above for the Groote Eylandt deposit, one wonders why this scenario was never repeated on a large scale in geologic history. It is

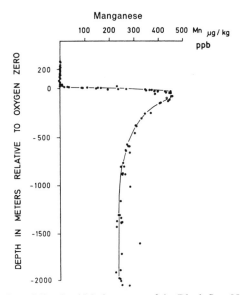

Fig. 5.10. Distribution of dissolved Mn in waters of the Black Sea. Note the precipitation of Mn at the redox interface. (Wedepohl 1969, fig. 25-I-1).

conceivable that the manganese was derived from the basin rather than from the rivers, as suggested by Borchert (1980). That is, the manganese could be dissolved from detrital sediments, then released to the overlying water as in Fig. 5-5, only in an isolated basin where the Mn would be deposited at the shoreline in more oxidizing water, rather than transferred to the deep-sea. The Black Sea today shows this behavior to an extent (Degens and Stoffers, 1977): both Fe and Mn are solubilized from oxides in the sediments by the reducing conditions in deeper parts of the basin, but the Fe is reprecipitated in sulfides, while the Mn is reprecipitated as an oxide at the boundary between oxidizing and reducing water (Fig. 5-10). At present, the shallow-water sediments are not strikingly enriched in Mn, but it is possible that a long-continued transgression could have suppressed terrigenous clastics while at the same time moving the zone of reducing conditions up to shallower depths, thereby leading to extensive Mn deposition near the shore line.

The carbonate-hosted deposits of Morocco probably get their Mn from weathering of older deposits, although this source does not seem likely to apply to all carbonate-hosted Mn. Further, the mechanism of precipitation is obscure.

It can be seen that we actually know little about ancient deposits compared with modern ones, and in this case most of the modern ones are not good analogues for the ancient. Much more work needs to be done on the petrography and geochemistry of individual deposits. As with many of the other ores described in this book, stable isotope studies should give a large return of information for the effort required. Another fruitful line of inquiry, particularly for the Precambrian deposits, is determination of the detailed facies patterns of the ore and its relation to Fe deposition.

CHAPTER 6.

Uranium

The Walrus and the Carpenter
Were walking close at hand:
They wept like anything to see
Such quantities of sand:
'If this were only cleared away,'
They said, 'it *would* be grand!'

Through the Looking Glass
From Lewis Carroll, *The Complete Works of Lewis
Carroll*. (New York: Vintage Books, 1976, page 184.)
Reprinted with permission of the publisher.

Unlike most other sedimentary ore metals, uranium forms small orebodies, which is in keeping with its low crustal abundance. As a consequence, it is a more exploration-intensive metal than the others described in this book. We have seen that deposits of Fe and Mn are vast, with reserves in single deposits as high as 10^9 tons of metal. For U, the largest deposits have 2×10^5 tons; the largest U.S. deposit (at Grants, New Mexico) has only 45,000 tons of U_3O_8 (Cheney 1981, table 4).

U accumulations occur in five distinct types that have a sequential distribution with time (Fig. 6-1). Oldest are the conglomerate Au-U deposits of the Witwatersrand, South Africa and Elliot Lake, Canada. The minerals are generally regarded as detrital, having accumulated under a low-oxygen atmosphere that permitted fluvial transport of grains such as pyrite and uraninite that are rapidly destroyed by oxidation in most modern streams. As detrital accumulations, these deposits are outside the purview of this book, but we will mention some features of their chemistry and petrography that differ from those of younger ores. Later in the Proterozoic is found an enigmatic type of accumulation, one whose economic importance has been appreciated only recently. These are the "vein" or "unconformity" type, found around the outcrop of the Athabasca sandstone in western Canada and in similar settings in rocks of about the same age in Australia. They are the largest and highest-grade deposits known (Cheney 1981). In the Phanerozoic, U mineralization occurs in black shales such as the Alum Shale of Scandinavia or the Chattanooga Shale of the eastern U.S., and in sandstones as the familiar "roll-front" and "tabular" deposits. The shale-hosted U is mostly of

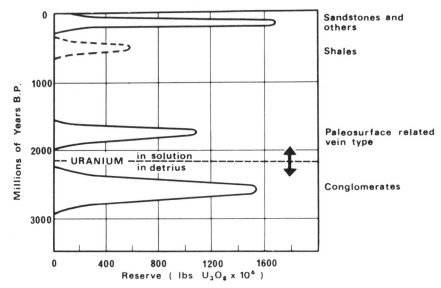

Fig. 6-1. Distribution of types of uranium deposits with time (Robertson and others 1978, fig. 2).

Paleozoic age, sandstone-hosted mostly Mesozic-Cenozoic. Sandstone deposits, which individually are fairly small, contain virtually all of the recoverable reserves in the U.S. The black shales have large tonnages, but the grade is so low, about 1/10 that of other types, that they have not yet been widely exploited. Finally, in the Tertiary to Recent, are the "calcrete" deposits of Australia and Southwest Africa. These surficial accumulations differ in having uranyl vanadates as the chief ore minerals.

Uranium, because of its low crustal abundance, is present in low concentrations in most rocks (Table 6-1). There is an enrichment in granites relative to other igneous rocks, and granites or rhyolitic ash are commonly cited as sources of the U in orebodies. Among sedimentary rocks, concentrations are also low, except in black shales, where U can be concentrated to ore grade. Most natural waters are extremely low in dissolved U, but note that the thorium/uranium ratio is lower than in most rocks, reflecting the greater mobility of U.

Mineralogy

A vast number of U minerals are known, but only a few are sufficiently common to be classed as ore minerals (Table 6-2). U minerals are almost always so small that microscopic techniques are of little use in their identification, resulting in a dearth of petrographic work on U ores. Identification is best carried out by X-ray diffraction on heavy-liquid separates.

Table 6-1. Abundance of U and Th in common rocks and natural waters.

	U, ppm	Th/U
Igneous Rocks:		
ultramafics	0.02	5.0
basalt	0.5	3.1
andesite	2	2.4
granite	4	4.9
Sedimentary Rocks:		
quartz arenites	0.45	3.8
graywackes	2.1	3.2
arkoses	1.5	3.3
shales:		
gray and green	3.2	4.9
red and yellow	2	6.5
black:		
average	53	
Chattanooga	79	
Alum Shale	168	
Ohio Shale	50	0.19
limestone	2.2	0.7
dolomite	1	
phosphorite	50–300	<0.1
Natural Waters:		
seawater	0.3–6	<0.03
ground water	0.3–10	
river water	0.03–10	<0.03

Source: From Wedepohl 1969, tables 90-E-1, 90-K-1, 90-I-1, 92-E-1, 92-K-1, 92-I-1 and Gabelman 1977, table 3.

Table 6-2. Composition of some common U minerals from ore deposits.

U^{4+} minerals:	
uraninite (more oxidized varieties termed pitchblende)	UO_2 to U_3O_8
coffinite	$USiO_4$
ningyoite	$CaU(PO_4)_2 \cdot 2H_2O$
U^{6+} minerals:	
carnotite	$K_2(UO_2)_2(VO_4)_2$
tyuyamunite	$Ca(UO_2)_2(VO_4)_2$
autunite	$(H,Na,K)_2(UO_2)_2(PO_4)_2$
uranophane	$Ca(UO_2)_2(SiO_3)_2(OH)_2$

Table 6-3. Representative analyses of the various types of U deposits.

	Percent					ppm			
	SiO_2	Al_2O_3	Fe_2O_3	MgO	CaO	Mo	U	V	Th/U
Conglomerate									
Au–U (Witwatersrand)	86.1	2.3	6.7	—	—	—	940	—	0.70
Unconformity vein (Rabbit Lake)	—	—	10	—	—	500	140,000	100–900	0.001
Black shale (Ohio Shale)	49.8	12.6	9.3	1.6	1.0	160	50	200	0.19
Roll-front (south Texas)	—	—	1	—	—	17	4000	—	—
Calcrete (Australia)	6.2	0.2	0.2	17.5	30.6	—	1700	320	—

Sources: Pretorius 1976, Tables 21, 23, 25; Knipping 1974, Table 2; Galloway and Kaiser 1980, Table 4; Mann and Deutscher 1978, Table 1).

Geochemistry

Some of the differences in the processes forming the various types of U ores are reflected in their bulk chemistry (Table 6-3). For instance, the Lower Proterozoic deposits are rich in thorium; calcrete deposits are high in V. Why this should be so will be explained later, but first let us examine the geochemical properties of U.

The behavior of U in aqueous solutions has been described by Langmuir (1978). It can occur in $4+$, $5+$, or $6+$ oxidation states, but the concentration of uranous (U^{4+}) species in solution is usually vanishingly small because of the insolubility of U^{4+} minerals. As a result, transport of U is in the $6+$ or possibly $5+$ state, and oxidation-reduction reactions are critical in the development of most U ores. Solubility, and hence transport, can be greatly enhanced by formation of complexes with other ions in solution. Figure 6-2 shows the speciation of U in the presence of the most common complexing agent, CO_3^{2-}. U^{6+}, as UO_2^{2+} also complexes with HPO_4^{2-}, OH^-, F^-, $H_2PO_4^-$, and SO_4^{2-}, in order of decreasing strength of association. By comparison, the strongest complexes for U^{4+} are OH^-, HPO_4^{2-}, F^-, and SO_4^{2-}. At the concentrations of these ligands found in natural waters, carbonate and phosphate complexes are the most important transporting agents, although fluoride can become important at low pH (Fig. 6-3).

Of the common uranium solids, uraninite is the most important. As shown in

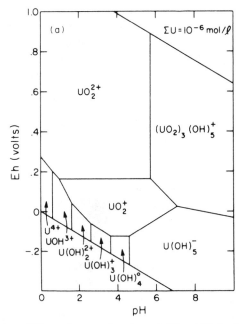

Fig. 6-2. Distribution of U species as a function of pH and Eh in the presence of CO_2. $P_{CO_2} = 10^{-2}$ atm, $\Sigma U = 10^{-6}$ (Langmuir 1978, fig. 2b).

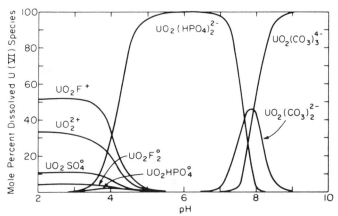

Fig. 6-3. Uranyl complexes vs. pH for ligand concentrations typical of ground water in the Wind River Formation of Wyoming. Note the predominance of phosphate and carbonate complexes in the neutral pH range. $P_{CO_2} = 10^{-2.5}$ atm, $\Sigma F = 0.3$ ppm, $\Sigma Cl = 10$ ppm, $\Sigma SO_4 = 100$ ppm, $\Sigma PO_4 = 0.1$ ppm, $\Sigma SiO_2 = 30$ ppm (Langmuir 1978, fig. 11).

Table 6-2, its composition ranges from UO_2 to U_3O_8. It is not known whether there is a stable solid solution between these end-members, or whether the intermediates are metastable. Langmuir (1978, p. 559-560) has suggested, by analogy with higher temperature relations, that a stable solution with cubic crystal structure exists between UO_2 and U_4O_9 and that the more oxidized compositions up to U_3O_8 are metastable (Fig. 6-4). Ludwig and Grauch (1980) report uraninite with the composition $U_{0.75}Ca_{0.15}Si_{0.17}O_2$. If this much Ca and Si commonly substitute in the uraninite structure, some of the phase relations depicted here may need to be modified. The other common 4+ mineral is coffinite, which was once believed to have OH in its structure, but is now regarded as a pure oxide. Again, it may have some Ca substituting for U (Ludwig and Grauch 1980, table 3). It is presumably stable relative to uraninite in silica-rich solutions, and, because both minerals are common, the silica concentration above which coffinite becomes the stable phase must be intermediate between average ground water (17 ppm) and saturation with respect to amorphous silica (about 140 ppm). In Fig. 6-5, 60 ppm (10^{-3} moles/l) was chosen.

Eh is the dominant control on U mineralization: low Eh leads to the precipitation of uraninite or coffinite, which are dissolved at high Eh (Fig. 6-6). Note also the sharp increase in U solubility resulting from carbonate complexing. The Eh response is much like that of Cu, and so U is also enriched by dissolution of dispersed metal by oxidizing solutions, followed by its precipitation at sites of reduction. Sorption of the 6+ species by organic matter or clays may generally precede this reduction, thus serving a catalytic or preconcentration function (Langmuir 1978, p. 558).

If appreciable amounts of V or phosphate are present, as may occur in arid climates when the parent rock is rich in these substances, uranyl deposits can form. The least soluble, and the most common, uranyl mineral is carnotite. Even

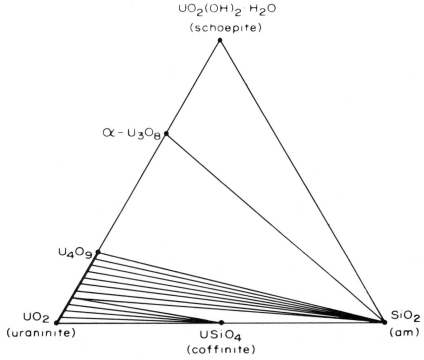

Fig 6.4. Possible phase relations in the system $UO_2 - SiO_2 - H_2O - O_2$ at 25°. The diagram suggests that coffinite should not be stable in the presence of more oxidized species such as U_4O_9 (Langmuir 1978, fig. 12).

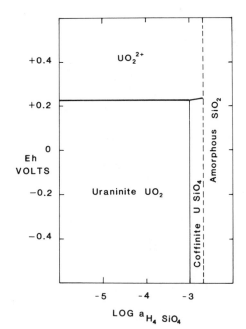

Fig. 6-5.
Suggested phase relations between uraninite and coffinite at 25°C. Both are stable only under reducing conditions.

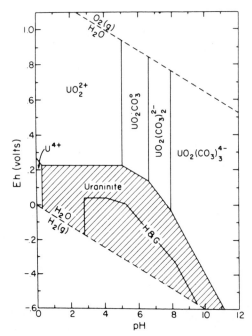

Fig. 6-6. Eh-pH relations for uraninite and aqueous solution with CO_2, demonstrating the necessity for reducing conditions for precipitation of uraninite at most pH's (Langmuir 1978, fig. 14).

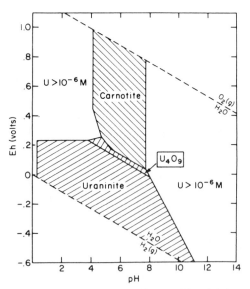

Fig. 6-7. Eh-pH relations under the same conditions as Fig. 6-6, but with V present. In this case, U can be precipitated under oxidizing conditions as well. $\Sigma U, V = 10^{-6}$, $\Sigma K = 10^{-3}$, $P_{CO_2} = 10^{-2}$ (Langmuir 1978, fig. 19).

small amounts of V in solution led to immobilization of U as carnotite under neutral to slightly acid conditions (Fig. 6-7). Notice that the uranium in this case is precipitated by oxidizing conditions, so that it would seem to be impossible to transport the U and the V in the same solution. If K is exceptionally low, it is possible for the Ca analogue, tyuyamunite, to form instead, or, if no V is present but phosphate is high, the autunites.

In summary, the geochemical behavior of U is controlled mostly by Eh, the amount of CO_2 in the system, and the concentration of V. Under reducing conditions, U is precipitated as uraninite or coffinite; under oxidizing conditions it is soluble unless V is present. Now, let us consider some specific examples of each of the five types of deposits to illutrate how this geochemical pattern is related to petrography, vertical sequence, and environment of deposition.

Examples of Types of Deposits

Early Proterozoic Au-U deposits

Large U deposits are found in quartz-pebble conglomerates of this age in many parts of the world, but the two main areas of production are from the Dominion and Witwatersrand reefs of South Africa and the Elliot Lake region of Canada (papers in Armstrong 1981; see also W.H. Gross 1968, Roscoe 1973, J.A. Robertson 1973, D.S. Robertson 1974, McMillan 1977, Pretorius 1981). Although the ores were originally believed to be hydrothermal, most workers now subscribe to a placer origin under conditions of low atmospheric oxygen. Exceptions are Kimberley and Dimroth (1976) who advocated a purely chemical origin, stressing similarities to younger U mineralizations, and Simpson and Bowles (1977, 1981) who, although accepting a placer origin, suggested that oxygen content of the atmosphere was not a factor because detrital uraninite grains are found in some modern streams in areas of rapid erosion. As shown in Fig. 6-1, however, ore deposits of this type have a striking predominance in the time period 2.8 to 2.3 $\times$ 10^9 years, which, especially in conjunction with the distribution of types of iron ores (Chapter 2), is most easily explained by a lower level of atmospheric oxygen.

The South African deposits have been exhaustively explored, and much is known about their sedimentology (Pretorius 1975, 1976, 1981, Minter 1976, 1981, Smith and Minter 1980, Viljoen and others 1980, von Backstrom 1981, Saager and others 1982b). Braided streams on alluvial fans or fan-deltas were the primary agents of deposition, with some modification by long-shore drift in distal areas (Fig. 6-8). The basin into which these deposits prograded was probably lacustrine or possibly an enclosed arm of the sea, but no unequivocally marine sediments are found. Like modern sediments of this type, these rocks consist of frequent fining-upward cycles, beginning with a quartz-pebble conglomerate and ending with very fine-grained material, often carbon-rich. Resistance to erosion makes the conglomerate layers stand out, and they are locally termed "reefs."

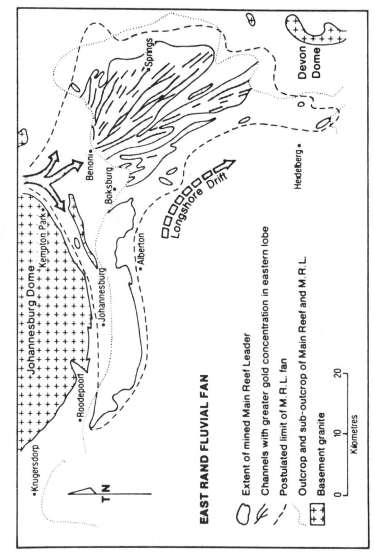

Fig. 6-8. Depositional setting of typical Witwatersrand Au-U deposit, showing a fluvial fan modified by long-shore drift, possibly in a large lake (Pretorius 1976, fig. 10).

The carbonaceous material is probably a remnant of algal mats that served to trap and bind detrital particles (Button 1979). Most of the mineralization is found close to the unconformities at the base of these cycles, and, accordingly, is associated in exploration with the conglomerate horizon. Genetically, however, it may be that the Au and U were originally deposited in the finer-grained part of the cycle, particularly with the carbon, and were then reworked up into the base of the next cycle by the erosion associated with the deposition of the gravel.

Hiemstra (1968), Feather and Koen (1975), and Schidlowski (1981) have described the petrography of the Dominion and Witwatersrand deposits; scanning electron photomicrographs can be found in Hallbauer (1981). In the Witwatersrand, three episodes of mineralization have been suggested: the first, a detrital phase during which gold, uraninite, pyrite, and cobaltite were deposited; the second, a diagenetic phase dominated by formation of pyrite, mostly as overgrowths on detrital cores with trace element concentrations different from the detrital pyrites (Saager 1981), and brannerite replacing muffin-shaped detrital uraninite (Schidlowski 1981, fig. 9); the third, a metamorphic or hydrothermal phase that led to extensive rearrangement of the gold on a local scale. Microprobe analyses of a number of the phases found are given in Table 6-4, and good color photomicrographs can be found in Feather and Koen (1975). The U occurs in two forms: as well-sorted detrital grains about 0.1 mm in diameter, and as very small (0.01 to 0.05 mm) euhedral crystals embedded in carbon. The size of the first type is hydraulically equivalent to that of associated zircon and chromite (Feather and Koen 1975, p. 199). Both varieties can be altered to brannerite, a titaniferous mineral believed to form at elevated temperatures. Further evidence for metamorphism comes from the composition of the arsenosulfides, which suggest temperatures of perhaps 400°C (Feather 1981, fig. 2). This metamorphism is also thought to have produced remobilization of the Au and, to a lesser extent, the U (Schidlowski 1968), although Hallbauer and Utter (1977, p. 299), based on similarity in morphology between Witwatersrand gold particles and those in modern placers, have argued that little remobilization took place.

The significance of the detrital pyrite and uraninite has been much debated. Most workers have subscribed to the view that atmospheric oxygen was absent or nearly absent (e.g., Schidlowski 1976), but Simpson and Bowles (1977, 1981), noting a similar assemblage in modern sands of the Indus River, proposed that oxygen content of the atmosphere was irrelevant. In this regard, the study by Hallbauer and Utter (1977) is especially revealing. They found that the gold particles in the Witwatersrand have retained their original Ag, but modern placer gold, transported similar distances, shows leaching of much of its Ag. Under oxidizing conditions, metallic Ag is converted to Ag^+ (Fig. 3-2). Thus, its preservation in the Witwatersrand gold particles implies an atmosphere that had considerably less oxygen than that of today.

An upper limit to the amount of oxygen in the atmosphere can be set by consideration of the kinetics of uraninite dissolution (Grandstaff 1974, 1976, 1980, 1981). The fraction of the original grain dissolved per unit time is a function of temperature, pH, dissolved carbonate species, surface area of the

Table 6-4. Microprobe compositions of phases found in Witwatersrand ores.

Arsenopyrite (%)		Cobaltite (allogenic) (%)		Gersdorffite (authigenic) (%)		Uraninite (%)		Brannerite (%)		Sphalerite (high iron) (%)		Sphalerite (low iron) (%)	
As	45.78	As	44.00	As	45.20	UO_2	66.8	UO_2	36.0	Zn	59.5	Zn	65.1
Ni	0.21	Ni	3.80	Ni	30.10	ThO_2	4.2	ThO_2	2.7	Fe	7.3	Fe	1.2
Co	0.17	Co	30.18	Co	2.40	PbO_2	23.8	PbO_2	11.7	S	33.0	S	33.0
Fe	32.30	Fe	2.14	Fe	2.60	FeO	0.7	FeO	4.1				
S	21.20	S	19.52	S	19.70	TiO_2	0.2	TiO_2	31.9				
						CaO	0.7	CaO	0.5				
								SiO_2	8.0				

Source: Feather and Koen 1975, table 4.

Table 6-5. Thorium content of some uraninites from placer deposits.

	UO_2/ThO_2	
Hunza River, Kashmir	8.5	
Indus River, Pakistan	9	
Indus River, Pakistan	15	
Dominion Reef	13.5	
Vaal Reef, Witwatersrand (allogenic)	19[1]	17[2]
Carbon Leader, Witwatersrand (fine pitchblende in carbon)	32[1]	22[2]

Source: [1] Simpson and Bowles 1977, fig. 10, 11.
[2] Feather 1981, table 4,5.

grains, their composition, and an empirical correction termed an "organic retardation factor," as well as the oxygen content of the water (Grandstaff 1980, p. 2-3). In order to estimate the maximum amount of oxygen that permits transport of appreciable amounts of uraninite, it is necessary to estimate the effect of these other variables. Values for most of them can be inferred from the modern or from experiments, but composition of the uraninite is important and must be determined on the original material. It turns out that impurities, especially thorium, retard the rate of dissolution, making thorian uraninites more resistant to weathering. From Table 6-5, it can be seen that the detrital uraninites contain considerably more Th than those that can be assumed to have been chemically precipitated. Also, note that the Dominion Reef ores have more Th than the slightly younger Witwatersrand. Robertson and others (1978, p. 1412) suggested that this difference might reflect an increase in atmospheric oxygen, but the high Th in the Indus River uraninites suggests that other factors may be responsible. Certainly more microprobe work is needed on uraninites of a variety of ages to see if there is an age trend. Using Th contents in the above range, Grandstaff (1980, p. 19) calculated oxygen contents for the Witwatersrand of 10^{-2} to 10^{-6} times that of the present atmosphere.

Late Proterozoic unconformity-vein deposits

This category of U deposits is the most controversial in terms of origin, yet has not been extensively studied petrographically or geochemically, and is presently the subject of much research. Therefore, any statements that we can make must be regarded as preliminary. These orebodies are found along unconformities separating highly folded, metamorphosed Archean and Lower Proterozoic rocks from essentially unmetamorphosed, relatively flat-lying Upper Proterozoic clastics (Fig. 6-9). Another distinctive feature is that the rocks beneath the unconformity are commonly deeply weathered. Rich orebodies have been found in Australia (Dodson and others 1974, Ayres and Eadington 1975, Hegge and Rowntree 1978, Eupene 1980) and in western Canada (Knipping 1974, Dahlkamp 1978, Hoeve and Sibbald 1978). Here we will discuss some aspects of the mineralization in Canada.

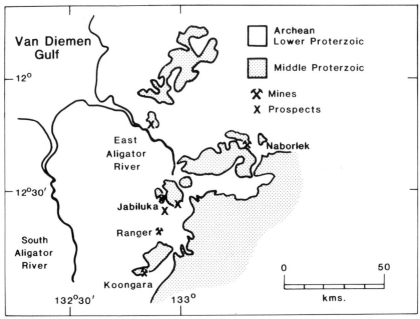

Fig. 6-9. Orebodies of unconformity-vein type in Australia lie close to the angular unconformity separating Upper Proterozoic from older rocks (redrawn from Hegge and Rowntree 1978, fig. 2).

There are recent descriptions of two of these deposits, Key Lake and Rabbit Lake (Fig. 6-10), which have a somewhat different style of mineralization. At Key Lake, Ni as well as U is mined (Dahlkamp 1978). Both are found at grades of one or two percent, and at least 50,000 tons of ore have been delineated. U minerals are pitchblende and coffinite. End-member uraninite (UO_2) is not found; instead there are two varieties of pitchblende : normal sooty pitchblende (UO_{2+x}) and tetragonal U_3O_7. This latter phase, in contrast to the sooty pitchblende, is visibly crystalline in reflected light, occurring as either small euhedra or radiating aggregates (Dahlkamp 1978, fig. 9-12). Note that if U_3O_7 proves to be a stable phase in this system, the relationships shown in Fig. 6-4 would have to be modified. The coffinite occurs as a replacement of the sooty pitchblende (Fig. 6-11). Ni minerals are gersdorffite (NiAsS), the most abundant, followed by millerite (NiS) and niccolite (NiAs), which is sometimes rhythmically intergrown with galena (PbS). Upwards within the orebody, crystalline U_3O_7 gives way to sooty pitchblende, and niccolite to millerite.

Most of the ore occurs in a strongly brecciated and mylonized fault zone that involves both the basement and the overlying Athabasca Sandstone. Stratigraphically, mineralization extends from about 20 m above the contact to perhaps 60 m below (Dahlkamp 1978, fig. 6). It is associated with kaolinization and, to a lesser extent, Fe-rich chlorite. Age dating of the U minerals gives a wide spread, from 1228-1160Ma for the crystalline U_3O_8 to 960-370Ma for the sooty pitchblende in the basement rocks, down to 200Ma for sooty pitchblende within the

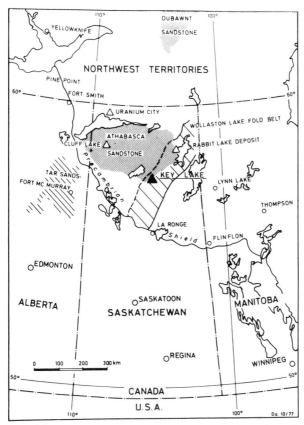

Fig. 6-10. Location of the Key Lake and Rabbit Lake deposits along the edge of the Middle Proterozoic Athabasca Sandstone, the same situation as found in Australia (Dahlkamp 1978, fig. 1).

Athabasca. Obviously, much redistribution of U has occurred, but all of the mineralization is younger than the Athabasca Sandstone, which was deposited before 1230Ma.

The Rabbit Lake deposits contain only uranium, which, like that at Key Lake, occurs in several generations (Knipping 1974, Hoeve and Sibbald 1978). Unlike at Key Lake, boron was present during mineralization, as shown by the presence of the Mg-tourmaline, dravite. Because the deposit has been well exposed by mining, the relations of the ore to the host rocks are reasonably well known, particularly the association of mineralization with alteration. At least three episodes of alteration occurred, and they affect the regolith and the Athabasca Sandstone, as well as the basement rocks, which indicates that the alteration, and hence the mineralization, post-date the Athabasca (Fig. 6-12). In the central part of the orebody, there is a limited area of dark green chloritization that is pre-ore. The next alteration was more pervasive, and comprised colorless (low Fe) chlorite with hematite and some dravite. Finally, there was a pale green

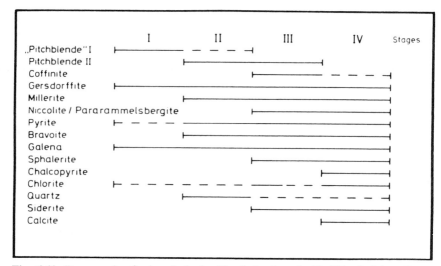

Fig. 6-11. Paragenesis of minerals in the Key Lake deposit. Pitchblende I is α-U_3O_7, pitchblende II is sooty pitchblende (Dahlkamp 1978, fig. 19).

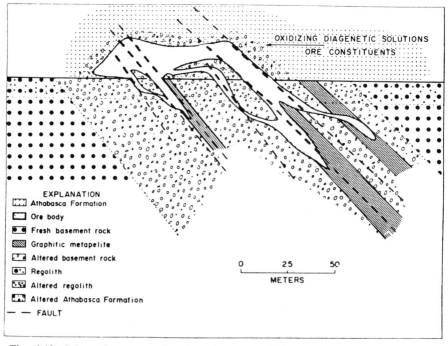

Fig. 6-12. Schematic cross-section through an unconformity-vein deposit showing the relationship between alteration, mineralization, and stratigraphy (Hoeve and Sibbald 1978, fig. 9).

Table 6-6. History of Rabbit Lake deposit.

1,350 ± 50 million years	—	Weathering of basement rocks; regolith
1,350 ± 50 million years	—	Athabasca Formation
	—	Brecciation
	—	Dark green chloritization (?)
1,100 million years	—	Stage 1 mineralization
	—	Brecciation–Rabbit Lake fault (?)
	—	Red alteration; chloritization; tourmalinization; silicification; dolomitization
	—	Brecciation
	—	Stage 2 mineralization; euhedral quartz veins
	—	Pale green alteration; some silicification; Stage 3 mineralization
	—	Younger reworking of deposit

Source: Hoeve and Sibbald 1978, table 2.

alteration, again with chlorite, but this time with small amounts of sulfides instead of hematite. Two episodes of mineralization were interspersed with these alterations (Table 6-6), the first a high reflectivity pitchblende in colloform encrustations, the second a more massive, low reflectivity pitchblende intergrown with coffinite and sulfides.

Rabbit Lake, Key Lake, and the Australian deposits share a position in fracture zones at the boundary between a sandstone and older metamorphics that, interestingly, contain graphitic schists. Hoeve and Sibbald (1978) have used this association to construct a ground water-epigenetic model for the origin of these ores. As shown by Table 6-5, the mineralization came somewhat later than the deposition of the Athabasca sands. Thus, it is reasonable to suppose that the U was deposited from circulating formation waters, somewhat in the style of roll-front deposits. If so, were these essentially cold meteoric waters from the surface, hydrothermal fluids from depth, or warm, deeply circulating ground water? And, what was the reductant responsible for precipitating the U? A study of the fabric of the Athabasca Sandstone in this region by Yu (1979) showed that the upper portion was probably buried to greater than 2400 m. Thus, the unconformity at its base must have been buried to at least this depth plus the 1750 m thickness of the Athabasca itself, or about 4200 m. Hoeve and Sibbald (1978, p. 1467) cite fluid inclusion data indicating a temperature of 220°C at 1.5 kbars and a salinity of 33 weight percent NaCl. At a geothermal gradient of 35°C/km, the burial depth would be 4800 m. Because the ore was emplaced shortly after the deposition of the Athabasca section, its conditions of formation must have been close to these. For temperatures of 200°C or higher, graphite will react with water to produce methane and some hydrogen. Hoeve and Sibbald hypothesized that this methane, migrating upward along fracture zones from the graphitic schists in the metamorphic basement, provided the reductant to precipitate the U which was being carried through the Athabasca by deeply circulating, oxidizing ground water (Fig. 6-13). Because of its slow reaction kinetics, methane may

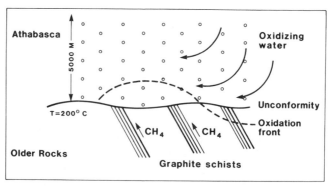

Fig. 6-13. Possible origin of the Athabasca deposits by interaction of organics from graphitic schists in the basement with U-bearing ground water.

not be an effective reductant, and other hydrocarbons, or reduced S species, might have been involved (Nash and others 1981, p. 86). Note that this process could be regarded as hydrothermal (it involves hot water), and Rich and others (1977) have classed these deposits as such. Furthermore, superficially similar vein deposits in Devonian granites in France are generally agreed to be hydrothermal (Cuney 1978, Leroy 1978). On the other hand, this model differs from most hydrothermal processes in involving oxidizing water moving down from the surface, more like a supergene process. Furthermore, the source of the uranium is from above rather than from the underlying metamorphics. Rich and others (1977, p. 64-72) discuss this and other possible combinations of oxidants, reductants, and sources of water. Needless to say, there have been many other hypotheses of origin proposed (e.g., Knipping 1974, Langford 1977, Dahlkamp 1978, Munday 1978, Eupene 1980). Some are just as plausible as the one just outlined, and only time will tell which proves to be the most satisfactory.

Black shale deposits

Next in the age sequence are some large, but low-grade, accumulations of U in black shale. None are commercial at this time, but two likely candidates are the Alum Shale (Cambrian) of Sweden and the Upper Devonian shales of the eastern U.S. Although both are Early Paleozoic, there seems to be no reason that similar shales of any age should not exist. For the U.S. deposits, the U will be a by-product of oil recovery, if it is ever produced, making these among the only low-grade ore deposits that will not require the consumption of large amounts of energy from off-site. In Sweden, considerable oil was produced between 1941 and 1965, and small-scale uranium extraction has gone on since 1953 (Martinsson 1974, p. 254-258). Plans had been made to mine a million tons per year as of 1979 (Frietsch and others 1979, p. 996), but environmental problems have delayed operation of the plants.

The Swedish shales are exceptionally rich in U, Mo, and V, with U_3O_8 contents ranging from 0.025 to 0.32 percent, averaging 0.035 percent (Armands 1972, Cheney 1981, p. 44). The mineable part of the unit is 2.5 to 4 m thick with recoverable reserves of 300,000 tons (Frietsch and others 1979, p. 996). The ores are contained in a thin sequence of shales and interbedded limestones believed to have been deposited in shallow water (Martinsson 1974). In the Early Cambrian, a transgressive sequence of sandstones and shales, the Laisvall Group, was deposited (Willden 1980, p. 96-98). Beginning with glacial-fluvial strata, the environments passed through transitional marine/non-marine to offshore marine by the beginning of the Middle Cambrian and the onset of Alum Shale deposition. Quiet, offshore marine conditions prevailed through the Ordovician, but with limestones becoming increasingly important. In contrast to the Devonian shales of the U.S., there is apparently no near-shore coarse-grained equivalent of the black shale facies. In places, it rests on the eroded basement (Bjørlykke and Englund 1979, fig. 1).

The Alum Shale section contains dark, bituminous shales, apparently devoid of trace fossils and bioturbation, with frequent beds and nodules of limestone. These are usually bituminous too, and hence are termed "stinkstones." Also found are small patches of coal-like organic matter referred to as kolm. The upper part of the shale is the richest in uranium, and generally U correlates with C (Fig. 6-14). Mineralogically, the Alum Shale contains only illite in the clay fraction, in contrast to the other shales in the section which have abundant chlorite; anomalously low Fe and Mn concentrations occur in the same interval (Bjørlykke

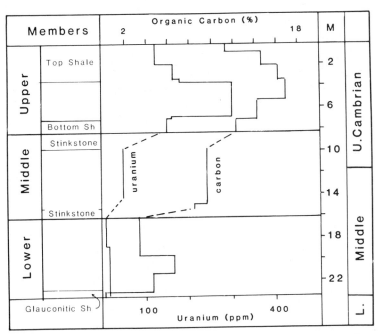

Fig. 6-14. Vertical section through the Alum Shale showing correspondence of uranium with carbon. Mo and Cr also reach values of 300 ppm in the uranium-rich interval below the Top Shale. (Data from Andersson and others, in press, fig. 3.)

and Englund 1979, fig. 4, 7, 10). Shallow-water conditions, probably just below wave base, are indicated by the presence of occasional beds of white, kerogen-free biocalcarenites composed of abraded brachiopod and trilobite fragments (Bergström 1980, p. 375). These must represent brief episodes of more agitated, better oxygenated conditions, perhaps related to a deeper storm wave base. Other parts of the Alum Shale have a low faunal diversity, although apparently the population density was high, consisting largely of olenid trilobites. In the Middle and Late Cambrian, these may have been planktonic (Bergström 1980, p. 376). It seems reasonable to conclude that the shales were deposited under at least mildly reducing conditions in water deeper than normal wave base. The rate of deposition was exceedingly slow, perhaps 1 mm/1000 yr (Bjørlykke and Englund 1979, p. 276), which may help to explain the high uranium: if the uranium is absorbed by the organic matter from seawater, as seems likely, then the longer the organics maintain contact with seawater, the higher the level of U and of other trace elements associated with carbon.

The Upper Devonian shales of the eastern U.S. have been studied intermittently for years as a low-grade source for U or oil (e.g., Conant and Swanson 1961). In the late 1970s, interest in these rocks as a source of natural gas prompted a re-investigation sponsored by the U.S.Department of Energy (1977, 1978, 1979) that has produced a wealth of new information. For uranium, the area of most interest is the western outcrop belt in Kentucky, Tennessee, and Ohio, where the sequence is thinnest. Here, the shales were deposited as the basinal facies of the westward-prograding Catskill delta (Rich 1951, Provo and others 1978, Potter and others 1982). The laminated, carbon-rich shales that make up the bulk of the distal part of the section were deposited under anaerobic conditions, in water deeper than the dysaerobic-anaerobic boundary (Byers 1977, Cluff 1980), as shown schematically in Fig. 6-15. They were also deposited very slowly, compared with the more proximal facies, as indicated by their sulfur isotopes (Maynard 1980). Along the crest of the Cincinnati Arch, the shale sequence is extremely thin, down to 5 m or less, and contains abundant phosphate nodules, suggesting that the Arch acted as a submarine high, much like the "Schwellen" of the Devonian in Germany. Similar uraniferous, organic-rich sediments are accumulating today in the Black Sea (Degens and others 1977), although in much deeper water.

U is high only in the organic-rich beds, in concentrations ranging from about 30 to 100 ppm, less than a third of that in the Alum Shale. Such values are far too low for recovery of U by conventional means, but pilot plant projects are now underway to produce hydrocarbons by mining and retorting the shale, and it is possible that U, and perhaps Mo, could be recovered as byproducts. In fact, it might prove necessary to remove the U from the tailings for environmental reasons. Table 6-7 shows an average of analyses for a typical sample in the area being considered for development. Note slight enrichment in Co, Cu, Mo, and Ni as well as U. Although the grade is very low, reserves are large. Total U_3O_8 in the shale in eastern Kentucky is about 180×10^6 tons (Provo 1977, table 6), but most of this would be inaccessible to surface mining. If the complete outcrop length of 500 km in Kentucky and adjacent Indiana and Ohio could be mined

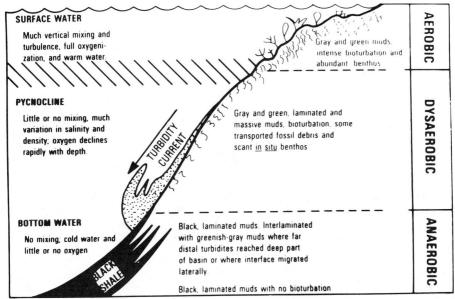

Fig. 6-15. Probable depositional setting of uraniferous black shales in the Devonian of the Appalachian Basin (Potter and others 1981, fig. 13).

over a width of 1 km, taking 15 m of shale whose average grade were 40 ppm, recoverable U would be on the order of 600,000 tons. Doubtless, the actual amount will prove to be much less because of incomplete extraction, unsuitability of some of the outcrop for strip mining, and low hydrocarbon recovery in some areas.

The form in which U is held in these rocks is not known, but the correlation between U and organic C (Fig. 6-16) suggests that it is somehow bound to the organic matter. Sorption onto the organic matter, rather than precipitation by

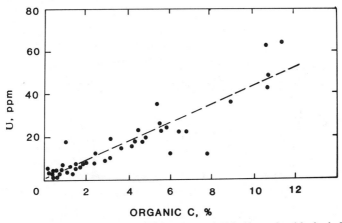

Fig. 6-16. Relationship between organic carbon and U in Devonian black shales of the Appalachian and Illinois Basins (after Leventhal 1980a, fig. 6).

Table 6-7. Chemical composition of black shale samples (SDO-1) from the base of the Ohio Shale (Devonian) of eastern Kentucky.

	Percent		ppm		ppm
SiO_2	49.8	As	80	Ni	100
Al_2O_3	12.6	Ba	440	Rb	140
Fe_2O_3	9.3	Cd	2	Sc	15
MgO	1.6	Ce	120	Ag	<1
CaO	1.0	Cr	70	Sm	50
K_2O	3.3	Co	50	Sr	100
Na_2O	0.42	Cu	67	U	50.4
TiO_2	0.67	Ga	19	Th	9.7
P_2O_5	0.12	La	50	Sn	400
MnO	0.045	Pb	30	V	200
		Li	30	Zn	60
C	10.5	Hg	<0.2	Zr	150
H	1.31	Mo	160		
N	0.35				
S	5.31				

Source: Average of determinations by various contractors of the U.S. Department of Energy, Eastern Gas Shales Project.

reducing conditions, seems to be involved because the low-organic beds can be rich in pyrite, but lack appreciable U. This organic material is both allochthonous, from terrestrial plant debris, and autochthonous, from planktonic algae or other plants within the basin (Maynard 1981). Thus, the proportion of each type and the duration of its exposure to seawater, from which the U presumably comes, may be important factors determining the concentration of U in the shale. Mo also correlates with organic matter (Fig. 6-17), but other elements such as Ni and V do not correlate with either C or S (Leventhal 1980a). Th tends to be fairly constant, so that the U/Th ratio also correlates with C (Leventhal and Goldhaber 1978). Another interesting relationship is between Mn and carbonate C (Fig. 6-18). In Chapter 5, we noted that, because Mn is usually mobile under reducing conditions, organic-rich shales tend to be depleted in Mn. In this case, Mn is low except in the presence of abundant carbonate, in agreement with the relationships shown in Fig. 5-2.

Sandstone-hosted deposits

Almost all of the U.S. production of U has come from small orebodies in Mesozoic-Cenozoic non-marine sandstones of Wyoming, Colorado, New Mexico, and Texas. As a result, deposits of this type have been studied intensively for years, and much is known about them. Commonly, they are of the roll-front type: ore occurs at the interface between oxidized and un-oxidized sandstone, in a crescentic or "roll" shape viewed in cross-section (Fig. 6-19). It is generally agreed that U^{6+}, carried by oxidizing ground water, is precipitated at this interface

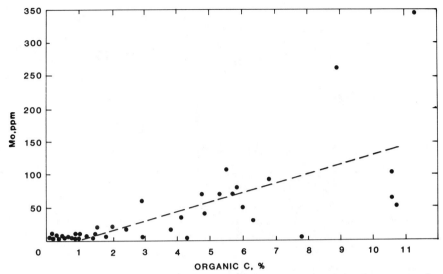

Fig. 6-17. Mo vs. C in the same samples as Fig. 6-16. The correlation is not as good, but there is an association between the two elements (after Leventhal 1980a, fig. 7).

by reduction to U^{4+}. The nature of the reductant, however, has been debated. Some workers argue for organic matter in the host sandstone (intrinsic reductant), others for an extrinsic source such as H_2S gas. Another variety, less easily explained, is the tabular type, which lacks the strong polarity of reduced and oxidized ground found in the roll-front deposits.

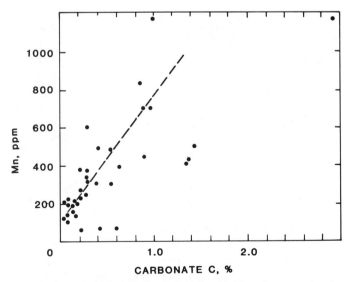

Fig. 6-18. Mn in Devonian black shales is associated with carbonate rather than organic carbon, suggesting incorporation into calcite and dolomite (after Leventhal 1980a, fig. 15).

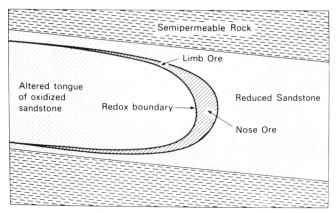

Fig. 6-19. Idealized cross-section through a roll-front deposit. The ore roll can be from 10s to 100s of meters from nose to tail (Evans 1980, fig. 15.8; after Granger and Warren 1969).

The ore minerals, where identifiable, are uraninite and coffinite (Ludwig and Grauch 1980) occurring as very fine-grained coatings on sand grains or as pore-fillings. U also appears to be adsorbed on clays. In addition to U, Mo and Se are commonly found, but not in commercial concentrations. Se normally is found close to the oxidation front, slightly updip from the U, whereas the Mo is displaced relatively far downdip (Fig. 6-20). This sequence is the same as the order of deposition with decreasing Eh (Harshman 1974, fig. 16; Howard 1977), and can be a valuable aid in determining the polarity of the roll-front if the strata have been re-reduced.

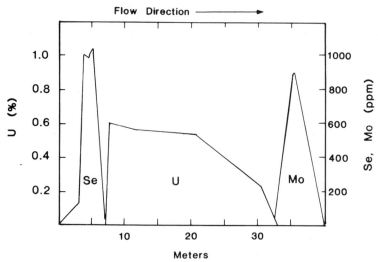

Fig. 6-20. Distribution of some elements with respect to the redox interface of a roll-front deposit. Note the sequence Se-U-Mo, with pyrite throughout. (Data of Harshman 1974, fig. 15.)

Many deposits, particularly in the Colorado Plateau region, do not have the roll-front shape, but are nearly concordant with the bedding and contain V and Cu as well as U (Fischer 1970). Genetically, they may be similar to the roll-front deposits (Rackley 1976), yet differ in having a more localized reductant such as patches of woody material or pyrite. Thus, instead of a redox front between updip (oxidizing) and downdip (reducing) conditions, there is a tabular region of reduction, parallel to bedding. There is evidence, however, from inferred patterns of ground water flow and from stable isotopes, that the ores were precipitated at the interface between meteoric water and basinal brines (Sanford 1982, Northrop and others 1982), in which case the reductant could be extrinsic. It is not known whether the appearance of V and Cu is a result of a difference in the details of the mineralization process or to a difference in the composition of the rocks supplying the uranium. A special case of tabular U is the large deposits of the Grants district, New Mexico, where the U is associated with amorphous-appearing organic matter that is thought to have been precipitated in the sandstones from ground water (Hilpert 1969, Adler 1974). Thus, both the organics and the U came in with ground water, and may have been introduced together or in alternating cycles (Leventhal 1980b).

To illustrate the principles outlined above, we will consider in more detail the deposits of the south Texas coastal plain. For more information on other areas, see the papers in the International Atomic Energy Agency Volume (1974) and by Rackley (1976), Adams and others (1978), and Nash and others (1981).

The geology of the south Texas deposits has been described by Eargle and Weeks (1973), Eargle and others (1975), Galloway (1977, 1978), and Galloway and Kaiser (1980). Predominantly fluvial sands and shales of Eocene to Pliocene age, interbedded with Eocene and Oligocene tuffs, dip seaward and are cut by numerous faults, usually downthrown in the seaward direction (Fig. 6-21). U mineralization occurred repeatedly, in as many as five stratigraphic units. These units show a strong volcanic influence: the shales are smectite-rich or even zeolitic; the sandstones are rich in plagioclase and volcanic fragments, both shards and intermediate to acid rock fragments. Within single units, the proportion of these volcanic grains relative to quartz decreases to the north (Galloway 1977, p. 14), suggesting a variation in provenance (Fig. 6-22). The uranium mines are concentrated in the southern part of the area, probably because of the higher proportion of volcanics in the section, although a more arid paleoclimate is another possible factor. The environment of deposition is also important, particularly the relationship of permeable and impermeable units. Commonly, deposits are located at facies boundaries that juxtapose rocks of different permeability. For example, in shallow deposits, the edges of crevasse splays, where the permeable sands finger out into impermeable overbank muds, are favored sites (Fig. 6-23).

The source of U in these, and in most similar deposits, is probably the interbedded volcanic ash, although granites have been favored by some (e.g., Rosholt and others 1971, Stuckless and others 1977) for the western U.S. deposits. For one of the important U-bearing units in this area, the Catahoula Formation, Galloway and Kaiser (1980, p. 14-15) found that paleosoil profiles contain 2 to

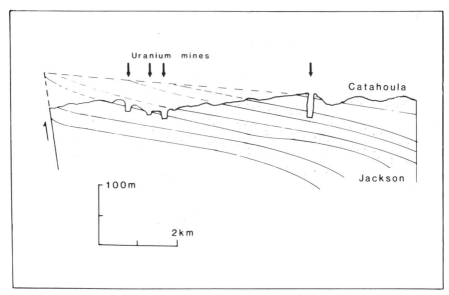

Fig. 6-21. Structure of U-bearing units in the south Texas coastal plain. U from the Catahoula was probably leached downward, then entered the truncated edges of the Jackson sands (redrawn from Eargle and Weeks 1973, fig. 4).

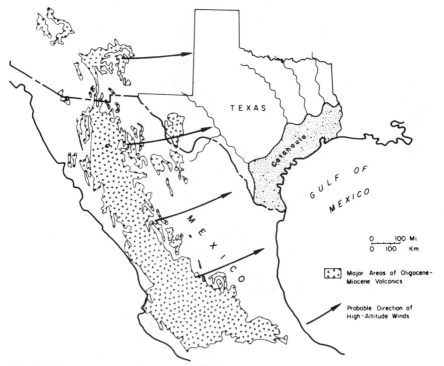

Fig. 6-22. Source areas of the volcanic debris in the south Texas coastal plain during the Tertiary (after Galloway 1977, fig. 3).

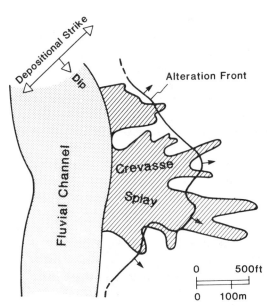

Fig. 6-23. U is concentrated in areas of permeability change such as the edges of crevasse splays (redrawn from Galloway and Kaiser 1980, fig. 28).

3 ppm U_3O_8, compared with 7 to 9 or more in unaltered material. Alterations of this type in volcanic-rich clastics seem to occur soon after deposition, mostly in the vadose zone (Walton and others 1981). In south Texas, evidence for vadose alteration includes clay coats on the sand grains, destruction of sedimentary structures by root disturbance, argillation of glass particles, and the formation of dispersed nodules of micrite. As might be expected, climate exerts an influence, as shown by the transformation of the glass to smectite in the southern, more arid part of the area, but to kaolinite in the north. After this early pedogenic phase, which was escaped by those parts of the formation deposited in lakes or swampy floodplains, a slower alteration in the phreatic zone occurred, distinguished by the formation of clinoptilolite and, in places, of sparry calcite cement. Smectite also formed, but appears as authigenic rims on the detrital particles rather than replacing glass. For U mobilization, the vadose stage of alteration is more important, as shown by the much stronger depletion of U in areas where it has occurred. Also, it has been found, in deposits of the western U.S. that, if tuffs are zeolitized, which characteristically occurs in the phreatic zone, they retain most of their U (Zielinski and others 1980).

The mobilized U enters shallow ground water circulation and is then precipitated by a reducing agent. For a few south Texas orebodies, the reductant was pods of organic matter, much as in the Colorado Plateau deposits, but with a clearly defined oxidation front. Galloway and Kaiser (1980, p. 49) have pointed out that if such a deposit were to be re-reduced, it would be classified as a tabular type. Most of the orebodies, however, are not associated with organic matter; the reductant was extrinsic, probably H_2S gas or aqueous sulfide species moving up along fault planes from hydrocarbon accumulations in the strata below.

How this leads to U enrichment has recently been elucidated (Reynolds and Goldhaber 1978, Goldhaber and others 1978). In the Benevides mine, the U is found in a typical roll-front at the boundary between oxidized and reduced sandstone. Almost no carbon is present in either portion, the difference being the predominance of iron oxides and hydroxides updip of the front, iron disulfides downdip. Close examination of the iron phases shows that the iron disulfides (pyrite and marcasite) replaced earlier Fe-Ti oxides, mostly titanomagnetite and, to a lesser extent, titanohematite. Immediately updip from the front, these sulfidized grains have been re-oxidized to produce a mixture of goethite and amorphous ferric hydroxide (limonite). Such grains are distinguished from detrital Fe oxides by their pseudomorphism of the sulfides and by the persistence of lamellae of TiO_2 in the trellis pattern of the original ilmenite. Farther updip, about 1 km, these grains give way to hematite and unaltered titano-magnetite (Fig. 6-24). Reynolds and Goldhaber (1978) concluded that the original sandstone contained only Fe-Ti grains of the type seen in the far updip area; that these were subsequently sulfidized by H_2S moving up the plane of a fault which is about 1.5 km downdip from the roll-front, the sulfidation being most intense near the fault and stopping at between 200 and 1000 m above the roll-front; and finally that the sulfidized portion of the sand body was oxidized by ground water moving downdip, which in the process deposited U at the redox interface (Fig. 6-25).

This scenario is confirmed by the mineralogy and isotope geochemistry of the sulfides. Roll-front U ores contain considerable quantities of iron disulfides (Fig. 6-20) that have isotopically light sulfur. Normally, such values suggest generation of H_2S by bacterial sulfate reduction, which may occur if the bacteria can use

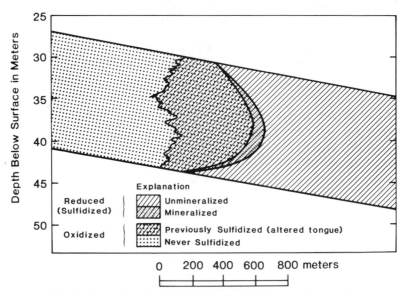

Fig. 6-24. Distribution of Fe-Ti oxides in the Benevides deposit showing patterns of oxidation of previously sulfidized grains (after Reynolds and Goldhaber 1978, fig. 4).

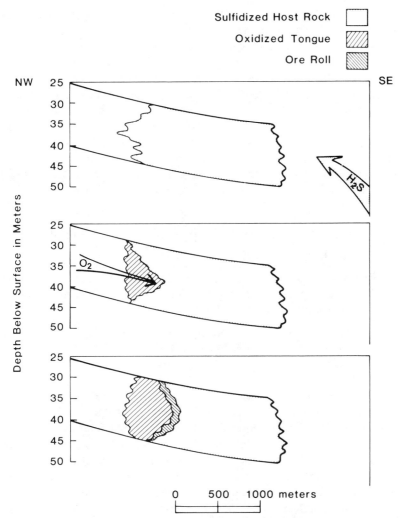

Fig. 6-25. Sequence of alteration and mineralization at Benevides. (A) H₂S moves up fault plane and sulfidizes Fe-Ti grains. (B) Oxygenated U-bearing ground water enters the aquifer, initiating roll-front, which (C) gradually moves down-dip concentrating U until the downward flow stops (after Goldhaber and others 1978, fig. 8).

the woody organic matter found in many of these deposits (Cheney and Jensen 1966). Alternatively, there are abiogenic reactions that produce the same effect that may be important in others cases (Granger and Warren 1969, Warren 1972, Cheney and Trammell 1973). Because of the lack of organic matter in most south Texas deposits (see Reynolds and others 1982 for an exception), their iron disulfides must have developed via the abiogenic route. In fact, there are two generations of iron disulfide, with different origins, mineralogy, and isotopic composition (Goldhaber and others 1978, Ludwig and others 1982). Downdip from the orebody, the sandstone contains only pyrite, which has isotopically heavy S, up to $+29$ permil, with an average of $+17$. Within the ore roll, marcasite is found as rims around pyrite, and is isotopically light, about -33 permil. Thus, there was a pre-ore sulfide, pyrite, that was deposited by the sulfidation of detrital Fe-Ti grains. The heavy S isotopes suggest derivation of the H_2S from crude oil, which in south Texas has S of about $+14$ permil. The marcasite is an ore-stage precipitate whose S was probably derived from metastable S species generated by oxidation of earlier-formed iron disulfides by the oxygenated ground water that introduced the U. That is, instead of going directly to the stable form, SO_4^{2-}, S from oxidizing sulfides may first go to metastable intermediates such as bisulfite, HSO_3^-, or thiosulfate, $S_2O_3^-$ (Granger and Warren 1969). For example,

$$8O_2 + 3FeS_2 + 6H_2O \rightarrow 2FeOOH + Fe^{2+} + 4H^+ + 6HSO_3^-.$$
$$\delta^{34}S: \quad +17 \qquad\qquad\qquad\qquad\qquad\qquad\qquad +17$$

On passing downstream through the roll-front, where conditions are reducing, some of the soluble Fe released in this oxidation step is reprecipitated as marcasite via disproportionation of the metastable S species:

$$Fe^{2+} + 7HSO_3^- \rightarrow FeS_2 + 5SO_4^{2-} + H_2O + 5H^+.$$
$$\delta^{34}S: \quad +17 \qquad -3 \quad +25$$

The acid released in these steps is probably responsible for the appearance of marcasite rather than pyrite. Experiments (Granger and Warren 1969, p. 169) show that in the disproportionation step, there is a strong preference for the light isotope by the reduced S, resulting, in one case, in delta values 23 permil lighter than the original material. In the south Texas deposits the change in delta is about 50 permil, indicating numerous cycles of oxidation and reprecipitation. For illustration, I have shown the isotopic values expected in the above reactions assuming a one-step fractionation of 20 permil.

To summarize, roll-front deposits, at least in south Texas, form in a sequence of steps involving sulfidation followed by oxidation. First, the host sandstone is "prepared" by the formation of pyrite from the reaction of fault-leaked H_2S with detrital Fe-Ti oxides. The upward movement of reductants is followed by downward movement of oxygenated ground water containing uranium leached from volcanic ash in the vadose zone. Reaction of this water with pre-ore pyrite or earlier ore-stage marcasite consumes the oxygen, thereby inducing precipitation of uraninite, and releases metastable S compounds from the sulfides, some of which go to form new marcasite, but most of which are lost to the system, resulting in a large isotope fractionation. The petrographic changes that accompany these reactions are shown schematically in Fig. 6-26. In some deposits, a later generation of pyrite forms in a re-reduction event (Ludwig and others 1982).

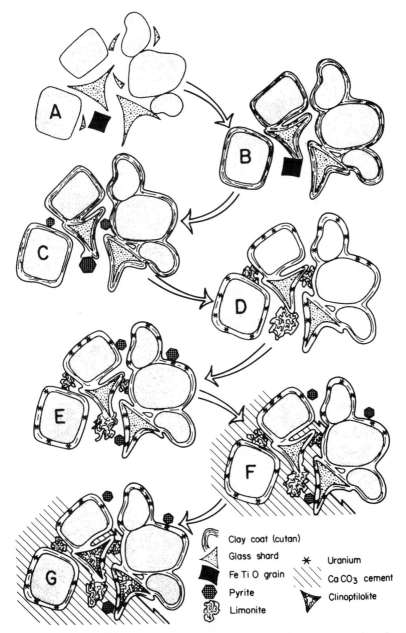

Fig. 6-26. Post-depositional petrographic changes in typical south Texas deposit. A. Initial detrital framework. B. Pedogenesis. C. Reduction and sulfidation. D. Oxidation and mineralization. E. Re-reduction. F. Calcite cementation. G. Zeolitization (Galloway and Kaiser 1980, fig. 27).

Calcrete deposits of Australia

The newest U deposits, both in age of formation and exploitation, are carnotite ores in calcareous surficial deposits of Western Australia. Little has been published about them. The account that follows is based on the paper by Mann and Deutscher (1978); some additional information can be found in Langford (1977). The carnotites are found in calcretes, which, in Australian usage, are deposits of Ca-Mg carbonates that precipitate from ground water near the surface. In this part of the continent, the drainage is mostly internal, with many playa lakes. The calcretes line the lower part of the drainage systems leading to the playas, where the water table approaches within 5 m of the surface.

Table 6-8. Chemical composition of calcrete U ore and nearby granite.

	Granite (3 samples)	Calcrete sample nos.	
		040045	040074
Percent			
Al_2O_3	13.79	0.21	0.18
SiO_2	73.60	23.31	6.20
TiO_2	0.19	0.01	<0.01
Fe_2O_3	1.69	0.14	0.17
MnO	0.04	<0.01	0.02
CaO	1.01	31.00	30.64
K_2O	4.47	0.08	0.09
MgO	0.32	9.45	17.45
P_2O_5	0.06	0.07	0.02
LOI	0.62	31.04	43.08
FeO	0.77	<0.05	<0.05
H_2O^+	0.11	1.13	0.73
H_2O^-	0.11	4.55	1.02
Na_2O	3.70	0.07	0.17
ppm			
Cr	6	<3	3
V	7	80	316
Ba	1050	569	<20
Nb	4	5	28
Zr	170	5	11
Y	10	1	<3
Sr	260	288	369
Rb	180	6	23
U	12	343	1693
Co	55	<2	2
Ni	20	28	27
Cu	3	11	3
Zn	50	30	10

Source: Mann and Deutscher 1978, table 1.

Chemical analyses of the rocks from one such basin show enrichment of the calcretes in U, V, and Sr (Table 6-8). It is believed that granitic rocks surrounding the drainage basin (Fig. 6-27) provide the U and Sr, but the source of the V is uncertain; it is high only in the immediate vicinity of the orebodies. Saturation of the ground water with respect to carnotite, expressed as the ratio of the measured to the equilibrium concentrations of its constituents, log IAP/K, increases towards the calcrete (Fig. 6-27), but only reaches super-saturation (IAP/K > 1) near the lake margin, suggesting that deposits "upstream" are presently being dissolved, while those at the edge of the lake are growing. The increase in saturation is attributable, in part, to the increase in potassium concentration caused by evaporation of the ground water, but also to a decrease in carbonate complexing of the UO_2^{2+} resulting from precipitation of the Ca-Mg carbonate of the calcrete. That is, total U stays about constant, but $a_{UO_2^{2+}}$ increases.

These deposits are distinctive in that the U is precipitated in the 6+ rather than the 4+ state. Therefore, oxidation-reduction reactions cannot be important for concentrating U in the ore, as is the case for roll-front deposits. Further, because the U is transported and deposited in the 6+ state, the V must have a different source, else no transport could take place. Alternatively, the V may be carried in the 4+ state in the ground water, to be precipitated as 5+ in the calcrete. In this case, oxidation rather than reduction would be the reaction responsible for ore formation. How this might occur is shown schematically in Fig. 6-28.

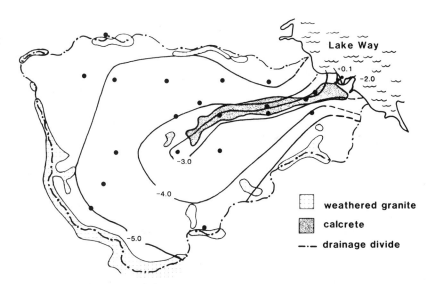

Fig. 6-27. Saturation state (Log IAP/K) of ground water with respect to carnotite in a typical calcrete drainage of Western Australia. The degree of saturation increases towards the calcrete, but exceeds one only at the lake margin (after Mann and Deutscher 1978, fig. 14).

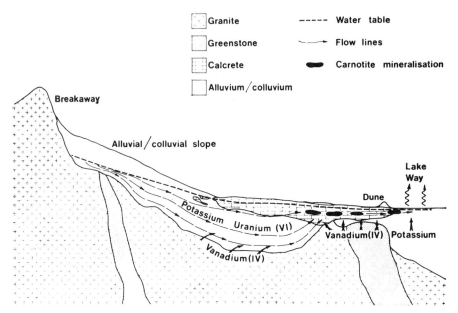

Fig. 6-28. Possible mechanism for the formation of calcrete deposits by oxidation of V^{+4} to V^{+5} combined with evaporation (Mann and Deutscher 1978, fig. 7).

Summary

U is found in a variety of deposits, ranging from detrital to supergene, that show a strong age dependence. Except for the detrital deposits, oxidation-reduction controls U behavior, with transport under oxidizing conditions, precipitaion at redox interfaces. The presence of abundant V, however, changes this pattern by favoring precipitation of uranyl vanadates in the most oxidized part of the system. As in oil exploration, the location of U deposits is governed by the presence of a suitable source rock with the correct "maturity," and a trap with the correct geometry. For U, the source is usually volcanic ash, leached soon after deposition; the trap is a porous sandstone or a fracture system that contains a reducing agent, which may be indigenous to the trap, or may have been introduced at some time prior to the entry of the U.

We seem to have a comparatively good understanding of the roll-front and quartz-pebble conglomerate deposits, but many uncertainties remain for vein deposits, and the study of calcrete deposits is in its infancy. For both of these types, the greatest need is for detailed descriptions of the mineralogy, geochemistry and geometry of further examples. For the roll-front deposits, we need to see to what extent the lessons learned from the south Texas deposits can be applied to those of Wyoming, and how both of these relate to the large deposit at Grants, New Mexico. Finally, we know as much about the thermodynamics of U deposition as we do about any other sedimentary ore metal, but the suggested importance of adsorption shows that some kinetic studies are needed as well.

CHAPTER 7.

Lead and Zinc

With that he roared aloud: the dreadful cry
Shakes earth and air and seas; the billows fly
Before the bellowing noise to distant Italy,
The neighboring Aetna trembling all around,
The winding caverns echo to the sound.

Polyphemus in *The Aenead*, Book III.
From John Dryden (trans.), *The Aeneid of Virgil*. (New
York: The Macmillan Company, 1965, page 114.)
Reprinted with permission of the publisher.

Lead and zinc are found in two distinct associations in sedimentary rocks: carbonate-hosted and shale-hosted. The carbonate-hosted ores are generally agreed to be epigenetic, deposited from low temperature hydrothermal fluids. The shale-hosted deposits are more controversial, but most appear to be at least partly syngenetic. Shale-hosted orebodies can be exceptionally large, and are perhaps the most popular exploration targets among sedimentary ores at the present time. A transitional form between these two, one that is accordingly given considerable attention, is syngenetic ore in carbonate rocks, such as at Tynagh, in Ireland.

A consideration of the distribution of Pb and Zn (Table 7-1) shows that Pb is uniform in sedimentary rocks, while Zn is somewhat enriched in carbonaceous shales. Both are remarkably low in carbonate rocks, suggesting that carbonate-hosted ores must be externally derived. Also, note the vanishingly small concentration of Pb in most natural waters, except highly saline ones. It would seem to be unlikely for Pb-Zn deposits to form by direct precipitation from seawater or other normal surface waters.

I. CARBONATE-HOSTED DEPOSITS

Mention sedimentary ores, and the first ones that come to mind for many geologists are the Pb-Zn ores so common in carbonate rocks. Their position does seem, often, to be controlled by sedimentary features, but there is good evidence that most are not sedimentary in the sense used in this book. Traditionally, most

Table 7-1. Concentration of Pb and Zn in some rocks and natural waters (ppm).

	Zn	Pb
Igneous Rocks		
peridotite	56	0.3
gabbro	100	3.2
diorite	70	5.8
granodiorite	52	15
granite	48	24
Sedimentary Rocks		
sandstone		
quartzose, arkoses	30	10
graywackes	95	20
shale		
average	100	—
carbon-rich	200	24
carbon-poor	—	23
carbonates	20	5
Modern Sediments		
marine mud	90	23
pelagic clays	140	55
Natural Waters		
seawater	0.005	0.00003
interstitial water	0.012	—
Salton Sea brine	780	80
deep formation brine, Canada	750	—
Atlantis II Deep brine	5.4	0.6
deep formation brines, Mississippi	155	30

Source: Wedepohl 1969, tables 30-E-8; 30-I-3; 30-K-2, 30-K-3, 30-K-4, 30-K-5; 82-I-1; 82-K-2, 82-K-3, 82-K-4.
Carpenter and others 1974, table 9.
Brewer and Spencer 1969, table 1.
Skinner and others 1967, table 1.

American workers have regarded these deposits as epigenetic, precipitated from relatively low-temperature hydrothermal solutions. Many European workers, on the other hand, persuaded by textural evidence, have maintained that they must be syngenetic. Sangster (1976) has proposed what seems to me an attractive resolution of this divergence: the North American deposits, which appear to always be void-filling, are, in fact, hydrothermal, while the Triassic ores of the Alps, which are commonly intimately interbedded with their host rocks, are partly epigenetic, partly syngenetic.

Mississippi Valley-Type Deposits

Lead-zinc ores in carbonate rocks occur in the U.S. in two broad areas: the Upper Mississippi Valley (Hagni 1976, 1982) and the folded Appalachians (Hoagland 1976). In Canada, such deposits are also common, as at Pine Point (Campbell 1967, Anderson and Macqueen 1982, Kyle 1982) and Robb Lake (Macqueen and Thompson 1978). A related group of deposits, in which fluorspar is the chief economic product, is found in southern Illinois and western Kentucky (Grogan and Bradbury 1967). Many carbonate-hosted ores have a close association with faulting (Fig. 7-1) that has, I think, inclined North American workers to view all such deposits as epigenetic.

The ores fill former open spaces that took the form of karst cavities, high porosity zones in bioherms, or faults. There is also an association between the ore and dolomitization. It is not known, however, how closely these two events are related in time: did the same solutions that dolomitized the limestones subsequently precipitate the ore minerals, or did the increase in porosity associated with dolomitization simply prepare favorable pathways for later, unrelated ore fluids? In some cases, such as the Tri-state area (Hagni 1976), mineralization is also associated with silicification of the limestone, but this is not true of the Appalachian ores, and so would not seem to be a critical genetic factor.

Based largely on data from fluid inclusion studies, which indicate a highly saline brine at 100-150°C (Roedder 1976, 1977), the ore fluid is generally believed to have been a chloride-rich formation water, probably expelled from basinal shales, much like the situation in the present Gulf Coast. Here, the shales are commonly undercompacted because of rapid deposition, giving rise to abnormally high fluid contents and pressures (Potter and others 1980, p. 240-247). Such waters may become enriched in heavy metals if they have a sufficiently high chloride content, perhaps from dissolution of evaporites, and the basin geometry

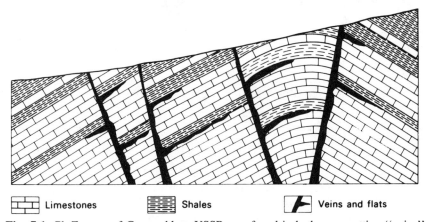

| Limestones | | Shales | | Veins and flats |

Fig. 7-1. Pb-Zn ores of Gyumushlug, USSR, are found in both cross-cutting ''veins'' and concordant ''flats'' (Evans 1980, fig. 2.9).

causes them to flow through metal-rich black shales. Carpenter and others (1974) have reported an example of one such basinal brine with 15 percent NaCl and temperatures of 90-160°C containing about 100 ppm Pb and 350 ppm Zn. Long and Angino (1982) have reported substantial leaching of metals from shales by brines: solutions with compositions similar to those of fluid inclusions in Mississippi Valley-type deposits removed as much as 10% of the Zn in several varieties of shales in 75 hours at 90°C. Evidence from the orebodies themselves for such a history comes from strontium isotopes (Kessen and others 1981). Most districts in the U.S. have gangue minerals with $^{87}Sr/^{86}Sr$ higher than the carbonate host rocks, indicating that Sr from silicate minerals has been added to the system.

The migration path for these fluids may have been provided by extensive unconformity surfaces that created high-permeability aquifers reaching down into the basin (Fig. 7-2); in other cases the presence of a permeable reef facies was probably important (Billings and others 1969). Deposition of the ore was probably caused by mixing with water from another source that contained abundant H_2S, possibly derived from petroleum (Fig. 7-3). Cooling of a single metal- and sulfur-rich fluid does not seem to be a feasible mechanism of deposition because the metals and reduced sulfur can only be transported together at these low temperatures if the pH is also low, incompatible with the coexistence of dolomite in these rocks (Anderson 1975, Beales 1975). Sverjensky (1981), however, has suggested that if CO_2 content of the brine is high, S and metals can be transported together. Direct evidence from fluid inclusions of the involvement of more than one fluid has been reported from one locality (Zimmerman and Kesler 1981),

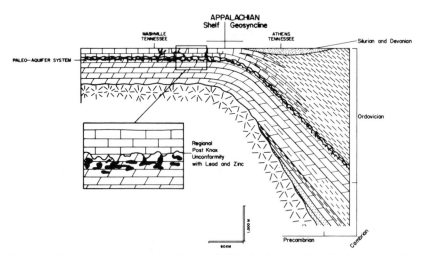

Fig. 7-2. Diagrammatic reconstruction of shelf-to-basin transition in the Appalachian Basin at the beginning of the Mississippian. The pervasive karstification below the unconformity at the top of the Knox may have provided pathways for basinal brines to migrate up-dip from overpressured shales, precipitating the Pb-Zn ores by mixing with meteoric water (redrawn from Hoagland 1976, fig. 20).

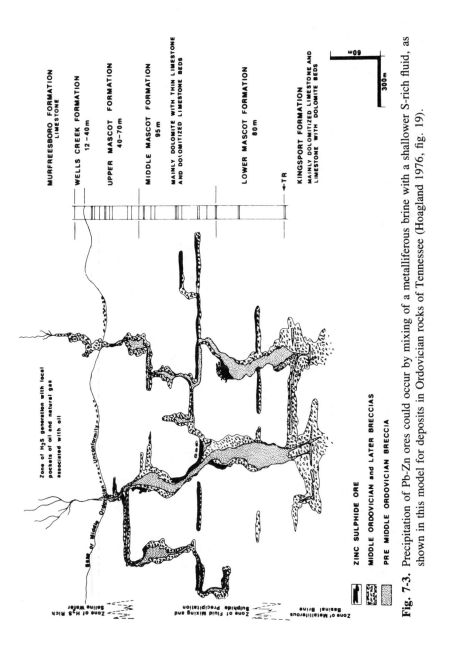

Fig. 7-3. Precipitation of Pb-Zn ores could occur by mixing of a metalliferous brine with a shallower S-rich fluid, as shown in this model for deposits in Ordovician rocks of Tennessee (Hoagland 1976, fig. 19).

and Sr isotopes can be interpreted as indicating two fluid systems (Kessen and others 1981).

Some additional features of the depositional process are revealed by the distribution of S isotopes. Compared with sulfides in modern sediments, the S is heavy, falling predominantly between 0 and +10 permil (Fig. 7-4), further evidence against a syngenetic origin. This range is, however, compatible with derivation of the S from petroleum or evaporites. Another difference between these and syngenetic sulfides is the attainment, in some deposits, of isotopic equilibrium between coexisting mineral pairs. For example, measurements on a sequence of ore minerals from a deposit in Wisconsin show a consistent spread of 4 permil between sphalerite and galena in early formed sulfides, but an abrupt shift to a larger difference in the later crystals (Fig. 7-5). Apparently, the S in the fluid from which the galena and sphalerite were precipitating changed from predominantly H_2S to HS^-, either because of decreasing temperature or increasing pH. The size of the spread between these two minerals is a function of temperature, as well as speciation of the S. Thus, if the form of S in the aqueous phase can be guessed, the isotopes can be used to infer paleotemperature (Fig. 7-6), although care must be taken to determine whether the mineral pairs were truly co-existing (Sverjensky 1981). In this case, the result is 227°C, higher than normally reported for such deposits, although MacQueen and Thompson (1978), using a variety of paleotemperature techniques, estimated temperatures of 200-230°C for ores in northeastern British Columbia. In summary, the overwhelming weight of evidence from petrography, isotopes, fluid inclusions, and field relations indicate that these ores are entirely epigenetic, having been deposited in the subsurface from saline brines at temperatures between about 100 and 200°C.

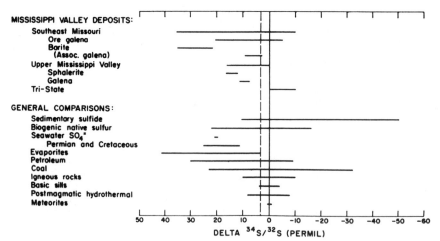

Fig. 7-4. S isotopes of Mississippi Valley-type deposits are generally heavier than those of sedimentary sulfides, by about 30 permil. Note, however, that the range is the same as that of reduced S in petroleum (Heyl and others 1974, fig. 6).

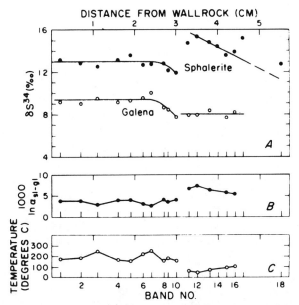

Fig. 7-5. S isotopes of coexisting galena and sphalerite can be used to infer the temperature and the state of the S in the precipitating fluid. In this case, the abrupt shift at 3 cm from the wall suggests a change from H_2S to HS^- in the ore fluid (Heyl and others 1974, fig. 7).

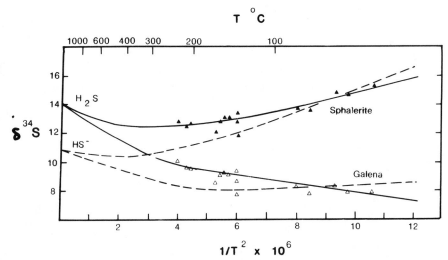

Fig. 7-6. Calculation of temperature from S isotopes. In this example, from the Booty Mine, Wisconsin, crystallization began at about 250°C and continued to below 100°C. Below 100°C the points fall on the HS^- (dashed) curve (after Heyl and others 1974, fig. 8).

Alpine Deposits

A different origin has commonly been ascribed to the Pb-Zn ores of the Alps. In places, these ores show a striking interbedding of ore minerals with shale and limestone (see especially Schulz 1964), plus a restriction to certain thin horizons of great lateral extent (Schneider and others 1977). Much has also been made over "geopetal" structures—those that indicate gravity settling of solid grains, preserving an indication of top and bottom, rather than precipitation from solution, where such polarity should be lacking. Examples are graded bedding and filling of one side of a cavity. A problem is that such textures in ores may be hard to distinguish from replacements. Furthermore, small accumulations formed epigenetically could subsequently by enriched to ore grade by supergene processes (Haynes and Mostaghel 1982), thereby giving the ore a sedimentary overprint. These observations have led to volcanic-syngenetic hypotheses (Amstutz 1959, Schneider 1964, Schulz 1964, Maucher and Schneider 1967, Pouit 1978), karst-syngenetic models (Bernard 1973, Collins and Smith 1973, Boni and Amstutz 1982, but see Schulz 1982), a reflux-dolomitization model (Geldsetzer 1973), a late dolomitization model (Macquar and Lagny 1981), and epigenetic hypotheses similar to those developed for the Mississippi Valley deposits (Duhovnik 1967, di Colbertaldo 1967). Brown (1970) provides an excellent summary of most of these ideas, and a detailed argument against syngenesis in the Mississippi Valley deposits.

The most popular hypothesis in the above group, at least judging from the number of papers advocating it, is volcanic-syngenesis. The Pb and Zn are thought to have been carried by hydrothermal solutions, ultimately of volcanic origin, and released onto the sea floor to be precipitated either by H_2S in bottom water or through sudden cooling of sulfide-rich hydrothermal fluids (Fig. 7-7 and 7-8). These primary sulfides are thought to have been remobilized during diagenesis and tectonism into the numerous cross-cutting forms now seen. Such a mixed epigenetic-syngenetic model is similar to that proposed by Dunham

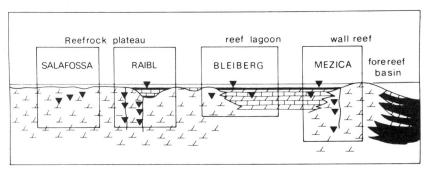

Fig. 7-7. Paleogeographic position of four major Alpine Pb-Zn deposits. Some are predominantly stratiform, others have considerable vein-like character (Brigo and others 1977, fig. 2).

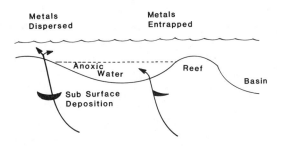

Fig. 7-8. Possible mode of emplacement of Triassic Pb-Zn deposits of the Alps. If fluids precipitate their minerals in the subsurface, an epigenetic deposit forms. If they reach the surface, a partly syngenetic deposit results.

(1964) for Pb and Zn in the Permian Marl Slate of Great Britain. Sulfur isotopes support an origin for these ores somewhat different from that of the Mississippi Valley-type: Table 7-2 shows the S to be lighter, consistent with derivation of at least some of the S from bacterial reduction of seawater sulfate.

Two important objections have been raised to this model: the scarcity of contemporaneous volcanics, and the small proportion of the ore that shows the syngenetic rather than the epigenetic features. These arguments are covered in detail in the discussions of the papers by Kostelka and Petrascheck (1967), Duhovnik (1967), and Schneider and others (1977). Possibly, volcanism is not required in this case; instead, the ore fluids are similar in origin to the Mississippi Valley-type—they just make it to the surface. By the same token, many of the vein-like Pb-Zn bodies may represent feeder conduits for these fluids, rather than diagenetic remobilizations. If so, we are seeing a transition from epigenetic to syngenetic mineralization.

Table 7-2. δ^{34}S of galena and sphalerite from carbonate-hosted Pb-Zn deposits.

	Mean	Range
Mississippi Valley Deposits		
Southeast Missouri	+12	±22
Upper Mississippi Valley	+ 7	± 7
Tri-State	− 5	± 5
Alpine Deposits		
Bleiberg	−20	±15
Mezica	−10	±10
Raibl	−18	±11

Source: Heyl and others 1974, fig. 6; Brigo and others 1977, table 1.

Irish Base-Metal Deposits

A more obvious epigenetic-syngenetic relationship is found in the Carboniferous Pb-Zn deposits of Ireland, especially at Tynagh (Derry and others 1965, Schulz 1966, Russell 1975, Boast and others 1981, Larter and others 1981). These ores differ from those of the Alpine region in having a more complex composition, with the addition of Ag and some Cu to the extractable metals, and a distal, tuffaceous Fe-Mn deposit, much like the volcanic-sedimentary deposits of the Kuroko type in Japan. Furthermore, there is a clear relationship with a fault, which was active during deposition of the host rocks. The geometry suggests that it was a conduit up which fluids moved to be mixed with seawater (Fig. 7-9).

Boast and others (1981) have presented a study that in many ways could serve as a model of how to examine sedimentary ore deposits. They combined detailed petrography with stable isotope analyses of S, C, and O, using samples carefully chosen for their place in the paragenetic sequence. The textures reveal four episodes of mineralization (Table 7-3): the first a diagenetic stage during which abundant pyrite formed, next geopetal precipitation of microcrystalline sphalerite and some galena in fractures in the host rock, followed by veining and replacement of the earlier-formed sulfides and the host carbonates by galena, barite, and tenantite, and finally formation of post-ore calcite veins.

Stage 1 sulfides are predominantly pyrite with minor marcasite and sphalerite occurring in highly convoluted bands or clots, about 1 or 2 cm across, in micritic carbonates of a Waulsortian mudbank complex. (See Wilson 1975, p. 148-169 for a discussion of this interesting carbonate facies.) On a small scale, the pyrite grains are seen to be discordant with laminations of the sediment, and the carbonate is recrystallized in the vicinity of the pyrite. Some laminae of microcrystalline sphalerite are found in pyritic horizons of shale beds interdigitated with the Waulsortian micrites, suggesting that some Zn as well as Fe was deposited syngenetically. For S isotopes in minerals of this stage, there is a range from -16 to -3 permil (Fig. 7-10), compared with a similar range of -22 to -9 for pyrite in limestone remote from the orebody. In both cases, the values indicate derivation of the S by reduction of SO_4^{2-} from seawater during open-system

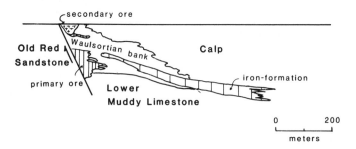

Fig. 7-9. Cross-section of the Tynagh orebody showing relationship of the iron formation and the sulfide orebody to the fault (after Russell 1975, fig. 1).

Table 7-3. Sequence of mineralization at Tynagh, according to Boast and others (1981).

Stage	Principal Minerals	$\delta^{34}S$ of Sulfides
1. Synsedimentary and early diagenetic	pyrite	-5.6
2. Geopetal deposition in fractures	sphalerite	-17.2
3. Veining and replacement of host carbonates and earlier sulfides	galena, barite, tennantite	-8.7
4. Post-ore veins	calcite	—

diagenesis at fairly low rates of sedimentation (Maynard 1980). Accordingly, the authors interpret these sulfides as normal products of early diagenesis. Most carbonate sediments, however, do not contain enough Fe to precipitate the amounts of pyrite they describe. Combined with the presence of the distal iron-formation, which is mostly hematite, this fact suggests that there was an addition of Fe from outside the system, probably from the fault. The S, on the other hand, could still be derived from bacterial reduction of SO_4^{2-}, provided sufficient organic matter were present.

Stage 2 minerals were deposited within a dilatant fracture system. Their textures are reminiscent of those reported from the Alpine Pb-Zn deposits, for instance interlamination of sulfides and host carbonate, graded bedding, and soft-sediment deformation features such as folds and miniature mud volcanoes. Close to the fault, the host limestone is brecciated, forming angular, poorly sorted blocks a few centimeters across in a matrix of microcrystalline sphalerite with minor galena. Dolomite and abundant barite were also precipitated at this stage. All of these phases are fine-grained, and settled by gravity following precipitation, rather than growing outward from fixed nuclei on the walls of the fractures, as in most vein deposits. They were probably precipitated rapidly within a highly supersaturated solution, from which they were then deposited much like clastic particles. Such high supersaturation implies a sudden change in some factor such as pH or temperature, compatible with mixing of a hydrothermal fluid with seawater. Sulfur isotopes are similar to those in Stage 1 (Fig. 7-10), except for barite, which is very positive, close to coeval seawater sulfate. The sphalerite is slightly lighter than the pyrite of Stage 1, but close to values for pyrite in the unmineralized limestone. Another indication that the sulfides at this stage were not receiving S from the fault is the lack of any relationship between $\delta^{34}S$ and distance from the fault. The authors conclude, as for Stage 1, that the S was derived from seawater, and, thus, that the ore was precipitated by the mixing of two solutions: one a metal-rich brine low in S, the other seawater enriched in reduced sulfur. The stability of dolomite at this stage shows that the metal-bearing fluid was not acidic, and therefore could not have carried appreciable sulfur (Anderson 1975).

Stage 3 marked a change from these conditions. Galena is predominant over sphalerite and tennatite appears along with lesser amounts of chalcopyrite, bornite and arsenopyrite; replacement textures are common; and S isotopes are much

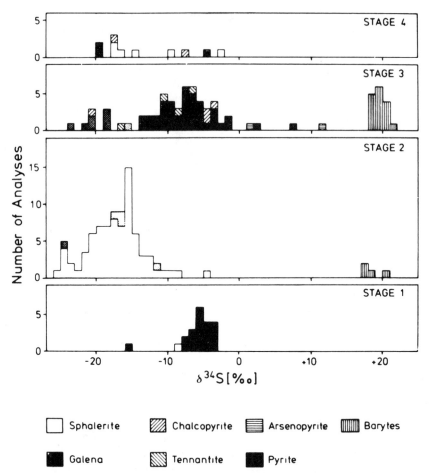

Fig. 7-10. S isotopes at Tynagh. Light values for sulfides and constant, heavy value for barite indicate a seawater source for S (Boast and others 1981, fig. 15).

heavier, especially close to the fault (Fig. 7-11). At Silvermines, another fault-related deposit some 40 km to the south, the same trend has been reported (Grieg and others 1971); see Taylor and Andrew (1978) for a discussion of faulting at Silvermines. Heavy S was probably introduced from the fault and mixed with light, bacteriogenic S. The source for the fault-derived S could have been deep-seated (mantle S is about 0 permil), but sulfides within the fault are too heavy, $+7.0$ to $+11.1$. Instead, it seems to me more likely that the source was again SO_4^{2-} in seawater, but this time abiogenically reduced at high temperature. Bachinski (1977, fig. 8) has calculated that pyrite precipitating in a basalt-hydrothermal system at 250°C should have $\delta^{34}S$ values between $+1$ and $+15$. Possibly Tynagh is a similar situation, but with a much thicker sedimentary section and more distant volcanic activity.

C and O isotopes were also measured. C becomes lighter in each stage of mineralization (Fig. 7-12), suggesting a mixture of mostly normal marine car-

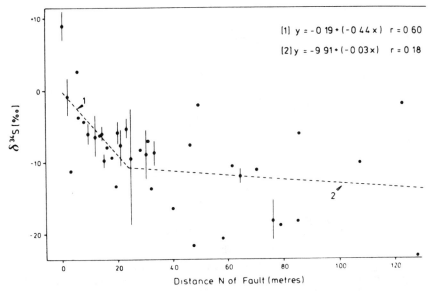

Fig. 7-11. For Stage 3 mineralization at Tynagh, S becomes lighter away from the fault, suggesting mixing of heavy S from the fault zone with light S produced diagenetically within the sediment (Boast and others 1981, fig. 15).

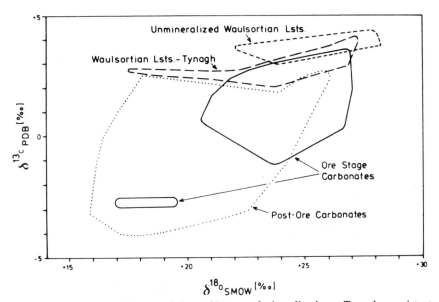

Fig. 7-12. Both C and O become lighter with stage of mineralization at Tynagh, consistent with addition of a small amount of organic carbon to normal sedimentary carbonate, and some interaction of heated seawater with wall-rock (Boast and others 1981, fig. 18).

bonate with small amounts of C derived from the oxidation of organic matter, which is much lighter. For O, the trends are irregular and suggest much disequilibrium. One quartz-albite pair from Stage 3 gave a paleotemperature of 200°C and $\delta^{18}O$ of the fluid of + 10.5. This compares with a temperature estimate of 200-250°C for Silvermines from fluid inclusions (Grieg and others 1971, p. 89). Such values are reasonable in the light of the sequence portrayed, but obviously much more work is needed.

For comparison with volcanic deposits, the Fe-Mn facies at Tynagh is particularly interesting. It has been described by Schultz (1966) who believed it to be unrelated to the sulfide mineralization, and by Russell (1975) who, like most workers, believed it to be a distal equivalent of the ores. The most abundant rock type is a hematite iron-formation composed of laminated to massive mixtures of hematite, quartz, and small amounts of other iron minerals such as Fe-chlorite (chamosite?), stilpnomelane, minnesotaite, and minor magnetic iron oxides, both magnetite and maghemite. Siderite is rare. Where laminated, the rock has a wrinkled appearance that Schultz attributed to "algal bedding." Chemically, the rock is similar to Precambrian iron-formation, being rich in Fe and Si and poor in Al and P (Table 7-4). The appearance of minnesotaite in the mineral assemblage is unusual in unmetamorphosed iron-formation, and needs further investigation. In the drill hole farthest from the fault, Mn as well as iron enrichment is found. The minerals are braunite and psilomelane in thin layers alternating with chert,

Table 7-4. Chemistry of iron formation at Tynagh compared with Precambrian BIF.

	Major Elements (%) Fe-rich Samples, Tynagh	Lake Superior Hematite IF
SiO_2	38.7	47.2
Al_2O_3	1.3	1.4
Fe_2O_3	43.2	35.4
FeO	4.4	8.2
MnO	0.1	0.46
MgO	0.4	1.2
CaO	4.2	1.6
TiO_2	0.1	0.03
P_2O_5	0.1	0.06
CO_2	6.1	—
H_2O	0.4	1.3

	Trace Elements (ppm) Fe-rich Samples, Tynagh	Mn-rich Samples Tynagh	BIF Lake Superior
Pb	170	1200	—
Zn	950	800	20
Cu	50	30	10
Ni	60	480	32
Ba	500	3830	180

Source: Russell 1975, table 1, 2; Gross 1980, table 3.

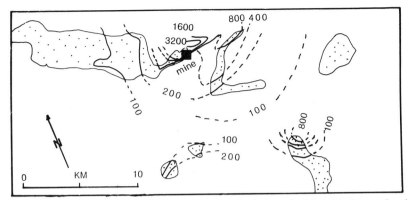

Fig. 7-13. Dispersion of Mn (ppm) around the Tynagh deposit. Many modern volcanic-sedimentary deposits show the same relationship (after Russell 1975, fig. 4).

and the rock is rich in Ba (Table 7-4). Both of these features, the Fe oxide to Mn oxide transition and the high Ba in the Mn oxides, are typical of volcanic-sedimentary deposits. Furthermore, there is a halo of Mn enrichment surrounding the orebody (Fig. 7-13), similar to many modern volcanic-sedimentary deposits (Cronan 1976; 1980, p. 278-282).

The foregoing suggests a history of successively hotter solutions moving up the fault to precipitate ore metals, both within the fracture zone adjacent to the fault and in the host sediments, contemporaneous with deposition (but see Evans 1976, who believed the ores to be post-depositional replacements). The metals released were first Fe and Mn, then Zn, then Pb and Ag. Ba accompanied the second and third episodes. The S was supplied by seawater, mostly by way of bacterial sulfate reduction, but with a lesser fraction derived from abiogenic reduction at higher temperatures. Some of the textures described are similar to the Alpine Pb-Zn deposits, and the facies pattern and mineralogy is reminiscent of Kuroko-type deposits (Chapter 8), except for the low Cu and lack of an immediate association with volcanoes. Thus, these ores may be part of a sequence: Mississippi Valley → Alpine Pb-Zn → Irish Base Metal → Kuroko in which the ores pass from epigenetic to largely syngenetic, temperatures increase, and composition becomes more complex. The source of the fluids and the S is ultimately seawater, either directly, in seawater-hydrothermal systems, or as connate brines; the source of the metals may be different in each case, possibly even within each type, as evidenced by the spread of Pb isotopes (Hoagland 1976, table 2).

Discussion

The hypotheses presented above explain many of the features of these deposits, but one particularly troublesome point remains: why is there such a close association with carbonates? Is it a pH effect, a neutralization of slightly acid solutions being the mechanism for precipitation? Or are the carbonates more efficient suppliers of reduced S? Does the association with carbonates somehow explain the near absence of Cu? For Tynagh, there is a similar problem: inspection of the geologic map presented by Grieg and others (1971, fig. 1) shows that all four commercial deposits of the Tynagh type in Ireland are found where Lower Carboniferous carbonates are in fault contact with the Old Red Sandstone, not where any of the other units in the section are on the upthrown side. This lithologic restriction is unexplained by the model I have presented.

The most important area for further work, it seems to me, is on ores of the Alpine type. Are they, in fact, somewhat similar to Tynagh, as I have implied? A comparable study of textures and isotopes should tell us. Perhaps more important, do any show the halo of Fe and Mn found at Tynagh? Finally, if comparisons are to be made with the North American deposits, more fluid inclusion studies, such as those reported in Pawłowska and Wedow (1980), will have to be undertaken.

II. CLASTIC-HOSTED DEPOSITS

Important Pb-Zn deposits are found in clastics as well as in carbonates. With the exception of Laisvall, in Sweden (Rickard and others 1979), all of the major clastic orebodies are in shales. Laisvall matches the Mississippi Valley-type deposits closely, being entirely epigenetic and deposited from a saline brine at about 150°C, but the major shale-hosted deposits share more features with the Tynagh style of mineralization, having a large syngenetic component to their deposition.

Two important areas of shale-hosted Pb-Zn are being mined and actively explored: western Canada and northern Australia. Because large new deposits are just being developed, an overall synthesis will not be attempted here; instead, I will describe one major deposit from each area: Sullivan, in British Columbia, and McArthur River in Australia. Other examples are Meggen and Rammelsberg in Germany (Anger and others 1966, Krebs 1972, Paul 1975, Hannak 1982, Krebs 1982), and various deposits of NW Canada (Thompson and Panteleyev 1976, Lydon and others 1979, Morganti 1981, Godwin and Sinclair 1982, Godwin and others 1982). Also see the abstracts of the symposium on shale-hosted deposits held at the 1981 Geological Society of America meeting, and the reviews by Large (1980) and Gustafson and Williams (1981). Possibly related to these ores are the massive pyrite orebodies discussed by Jenks (1975).

Sullivan

Several deposits of Pb and Zn, but minor Cu, occur in the Proterozoic of British Columbia, Canada, the largest being Sullivan (Ethier and others 1976). Like Tynagh, there is an association with a fault (Fig. 7-14), and the distal portion of the orebody consists of bedded sulfides, apparently syngenetic. Instead of

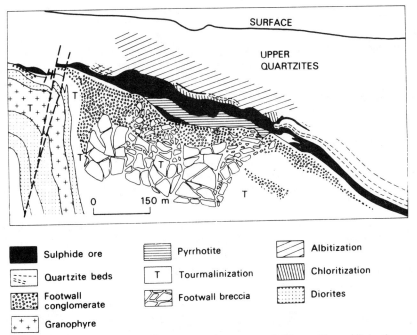

Fig. 7-14. Cross-section through the main ore zone at Sullivan. Tourmalinization and brecciation are confined to the footwall (after Evans 1980, fig. 2.11).

carbonates, the host lithology is interbedded shales and turbidite siltstones. Again, volcanics are absent. Sullivan, however, has been metamorphosed to at least chlorite grade (Ethier and others 1976), obscuring some petrographic and geochemical features. Unlike Tynagh, there is an extensive zone of alteration beneath the orebody, extending a short distance up into the hangingwall. Instead of a mineralized fault breccia, there is a synsedimentary, tourmalinized breccia and conglomerate that forms the footwall of the orebody, indicating that the ore-forming fluids contained large amounts of boron. Hangingwall alteration shows no tourmalinization; instead albitization is the predominant reaction. The ore minerals are mostly galena and sphalerite with minor chalcopyrite and, oddly, some cassiterite. Iron sulfides are represented by abundant pyrrhotite as well as pyrite. Thus, at least three events occurred during and shortly after the deposition of the host rocks: brecciation and introduction of boron; introduction of Fe, Pb, and Zn; and finally Na metasomatism.

It seems likely that the metals were carried by a chloride-rich brine low in sulfur. A sulfur-rich hydrothermal fluid should have led to mineralization of the underlying breccia. Precipitation of the metal sulfides may have been caused by mixing of the brine with H_2S-laden bottom waters in an anoxic basin (Campbell and others 1978, p. 267). A similar situation could arise today by a slight restriction of circulation in the Red Sea: the ensuing anoxia combined with the influx of metal-rich brines would produce orebodies much like Sullivan, only with a close volcanic association (see Chapter 8). Significantly, Sullivan also appears to have formed in a rift (Harrison 1972, fig. 2; Sawkins 1976b, p. 666; Badham 1981, p. B72).

The sulfur isotope data, however, are ambiguous regarding this model. Campbell and others (1978) report a range of -12 to $+4$ permil, with a slight trend towards lighter values upward (Fig. 7-15). Galena and sphalerite show similar values, with some reversals, indicating failure to equilibrate. Both features are consistent with a mixing model, but the average $\delta^{34}S$, -2.2 permil, seems heavy compared with younger deposits, if all of the sulfur was supplied by bacterial reduction of seawater sulfate. On the other hand, because we have so little information with which to reconstruct seawater compositions for the Proterozoic, this value may still be consistent with a seawater source.

A peculiarity of Sullivan that needs explanation is the predominance of pyrrhotite, whereas pyrite (plus marcasite) is the only iron sulfide in almost all sedimentary rocks and sedimentary ores. Conceivably, the massive pyrrhotite in the central part of the orebody at Sullivan was deposited at a relatively high temperature. Pyrrhotite also occurs widely dispersed in turbidite siltstones of this age in the Purcell-Belt basin (Huebschman 1973). There is a strong possibility that this pyrrhotite is detrital, being derived from centers of deposition like Sullivan. If so, there may also be detrital accumulations of Pb and Zn, analogous to those found in the down-slope direction from the *in situ* orebodies at Buchans (Calhoun and Hutchinson 1981). Other examples of accumulation of sulfides as turbidites have been reported by Skinner (1958) and by Schermerhorn (1971).

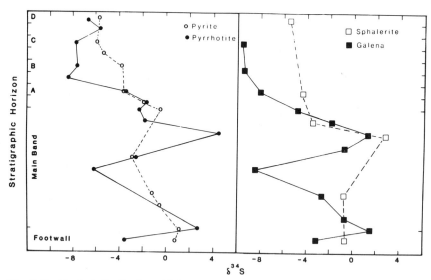

Fig. 7-15. Vertical distribution of S-isotopes in the eastern portion of the Sullivan orebody by stratigraphic horizon. Values become more negative upwards; note the lack of equilibration between galena and sphalerite. Thickness of the section varies between 50 and 70 m. (Data of Campbell and others 1978, fig. 4.)

McArthur River

Even larger than Sullivan are the deposits of the McArthur River area of northern Australia (Murray 1975, Lambert 1976, Williams 1978a, b, Lambert 1982). They are particularly important because they are unmetamorphosed, and seem to serve as analogues for the more easily mined deposits at Mt. Isa and Broken Hill, where metamorphism has obscured many of the depositional features, but has increased the grain size of the sulfides sufficiently for easy separation (Lambert 1976, p. 572-578). For information on these other ore deposits see the references cited by Lambert and also Stanton (1976). At McArthur River, most of the mineralization is found in stratiform ores of the H.Y.C. ("Here's Your Chance") orebody, but there is also some discordant mineralization to the east, associated with the Emu fault. The distribution of these two styles of mineralization resembles that at Tynagh (Fig. 7-16), but with the relative amounts of concordant and discordant mineralization reversed. Close to the fault, there is mineralized breccia with textural evidence of open-space filling, while the stratiform ores are delicately interlaminated sulfides, dolomite, and carbonaceous shale, sometimes tuffaceous.

In the H.Y.C. orebody, the sequence has been divided into 8 orebeds with 6 sub-economic inter-ore beds (Murray 1975). The ore beds are carbonaceous, tuffaceous shales with laminae, usually monomineralic, of extremely fine-grained metal sulfides. The tuffaceous component is exceptionally rich in potassium (Lambert 1976, p. 542), some of which may have been added to the sediment during diagenesis. The sulfide-rich laminae show abundant evidence of penecontemporaneous deformation, particularly slumping, which indicates that the sulfides were emplaced prior to lithification of the sediments. Within a given ore bed, there is a lateral zoning of the metals (Murray 1975, p. 336) in the sequence Pb-Zn-Fe. For the lower beds, this is the trend radially outwards from the center of the basin, but for higher beds, the richer ore zone is shifted eastwards towards the Cooley Dolomite Member. The inter-ore beds comprise mostly dolomite turbidites and breccia beds. They thicken eastwards, towards the inferred carbonate bank represented by the Cooley Dolomite (Fig. 7-16), and are thought to be a fore-reef facies of the Cooley.

There is a zoning of metals in the district as well as within single beds (Fig. 7-17), apparently related to the Emu Fault. As at Tynagh, Cu is high in the immediate vicinity of the fault, while Pb and Zn predominate in the distal direction. Other elements whose distribution appears to be related to the H.Y.C. orebody are manganese, arsenic, mercury, and thallium, all of which are relatively high in shales and dolomites laterally equivalent to the deposit (Lambert 1976, p. 559-561). The Mn dispersion halo is reminiscent of that at Tynagh and at Meggen (Gwosdz and Krebs 1977). The Cu/(Pb + Zn) distribution suggests that metal-bearing solutions came up the fault zone, then discharged onto the sea floor. If they were saline brines more dense than seawater, they would have moved downslope to accumulate in topographic depressions (Turner and Gus-

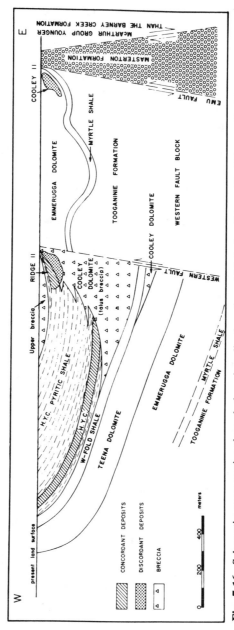

Fig. 7-16. Schematic cross-section through the mineralized section at McArthur River. The Western Fault appears to have been active during deposition, with carbonate bank deposits on the upthrown side and deeper-water talus and shale forming on the downthrown side (Williams 1978, fig. 4).

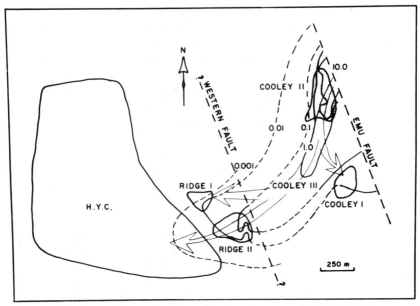

Fig. 7-17. Plan view of McArthur River deposits. Contours of Cu/(Zn + Pb) suggest a relationship to the Emu Fault. Arrows show inferred direction of fluid flow, which may have been within the carbonates in the eastern portion, but emerged near the Western Fault to flow into the H.Y.C. sub-basin at the surface (Morganti 1981, fig. 4).

tafson 1978, Solomon and Walshe 1979). Alternatively, if they were at a relatively high temperature, they would have been less dense than seawater, despite their higher salinity, and would have been bouyant. In shallow water, this bouyant plume would spread out at the water surface, raining down fine grains of galena and sphalerite, while the Cu was being deposited in and near the vent, a model that Solomon and Walshe (1979) suggest for the thin, sheetlike volcano-sedimentary deposits at Rosebery in Tasmania. In a deeper-water basin, they could be completely dispersed.

In addition to the question of the density of the ore fluids, there is that of the source of the sulfur. At least four scenarios can be envisioned:

Ore fluid more dense than seawater:

1. carries Zn, Pb, and S into depressions;
2. carries Zn, Pb only, S derived from seawater.

Ore fluid less dense than seawater, carries metals and S:

3. sulfides precipitate from bouyant plume over basin;
4. sulfides precipitate at vent, redistributed by turbidity currents.

Williams (1978b) has argued against the second possibility on the basis of S/C ratios. In modern sediments, reduced sulfur is positively correlated with organic carbon with a relatively constant ratio of about 0.10-0.15 (Williams

1978b, fig. 2), but in the McArthur River orebodies, S/C greatly exceeds these amounts, indicating a source other than bacterial reduction of seawater sulfate for much of the S. Williams then went on to propose an epigenetic origin for all of the orebodies, including the H.Y.C., but this seems unlikely from the textural evidence (Croxford and others 1975, Lambert 1976, p. 565-567).

S isotope data can also be used to help decide among these possibilities. Smith and Croxford (1973, 1975) report different isotope patterns for S in galena/ sphalerite and in pyrite, suggesting two sources for the S. Pyrite shows a trend similar to that in the Kupferschiefer (Chapter 3), with S becoming heavier upward (Fig. 7-18). Consideration of the hangingwall and footwall pyrites suggests that this trend is the middle of three cycles of increasing $\delta^{34}S$. Each appears to begin with an abrupt shift to light values, followed by a gradual increase upward of 15 to 20 permil. As in the Kupferschiefer, these trends can be explained by a gradual depletion of the sulfate of the water in a restricted basin of deposition, leading to progressive enrichment of the remaining sulfate in the heavy isotope. The sudden lightening of the S at the base of each cycle would then reflect renewal of free communication of the basin with the oceans. Deposition in a restricted basin is supported by casts and pseudomorphs of halite and gypsum in the dolomites (Lambert 1976, p. 538). Galena and sphalerite do not vary

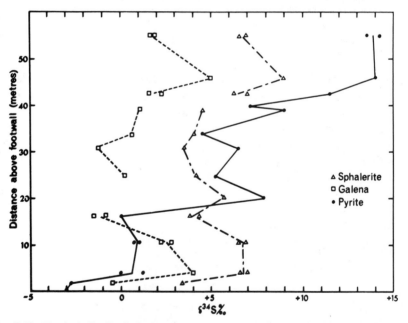

Fig. 7-18. Vertical distribution of S-isotopes in the H.Y.C. deposit at McArthur River (Lambert 1976, fig. 13; after Smith and Croxford 1973). Pyrite immediately below the ore zone averages +13.5 (Smith and Croxford 1975, table 1) showing a discontinuity in deposition. Similarly, pyrite in the hangingwall beds shows a return to lighter sulfur (+ 0.16) from the very positive values at the top of the ore zone.

sytematically in S isotopic composition vertically. They are separated by a nearly constant difference of about 4 permil, which, if they were deposited in isotopic equilibrium, would indicate a temperature of precipitation between 170 and 220°C (Smith and Croxford 1973, p. 11). Because of this consistency in their isotopes, plus their very fine intergrowths, galena and sphalerite must have been formed together from the same fluid. From their exceedingly small size (<10 micrometers), they must have precipitated rapidly. Furthermore, the difference in isotopic behavior from pyrite suggests that their S did not come directly from seawater, whereas for pyrite, a seawater source for at least some of the S is likely. It may be that the Fe as well as the Pb and Zn was exhalative, being deposited first as FeS, which was converted to FeS_2 by reaction with S from seawater. Alternatively, the Fe could have come from within the basin, a choice supported by differences between the isotopes of lead in the pyrite and in the galena/sphalerite (Gulson 1975). Pb in pyrite is much more variable and more radiogenic than Pb in the other two sulfides. Moreover, it is similar to Pb found in acetic acid washes of the dolomites, indicating that at least a portion of the Fe in the deposits was not exhalative. This analysis suggests that the Pb, Zn, and S were carried by the same fluid, then the Pb-Zn sulfides precipitated rapidly near the vent, and, along with dolomite grains, were redistributed by turbidity currents (alternative 4 above). But note that Williams and Rye (1974) preferred an epigenetic interpretation of this same S data, and also used the trend of C and O isotopes of mineralized vs. unmineralized dolomites to argue that the beds had been exposed to hydrothermal fluids after deposition (Rye and Williams 1981).

Discussion

We need to consider the source of the ore metals for these deposits. It is generally agreed that normal seawater does not contain high enough concentrations of metals (e.g., Rickard 1973), thus there must be some unusual concentration mechanism. By analogy with the Mississippi Valley-type deposits, and also those of the Red Sea (Chapter 8), a chloride-rich brine is often invoked. Debate centers on whether this brine was heated volcanically (Lambert 1976) or was a normal connate brine expelled from a compacting sediment mass (Badham 1981, Morganti 1981). The expulsion of formation waters is a particularly attractive mechanism because it explains large Pb-Zn deposits in terrains with minor volcanic material. How do these waters obtain their metals and how are they injected into a sedimentary basin? One explanation for the metals is that they are derived from a metal-rich source rock, probably a bituminous shale, analogous to a petroleum source bed. Chloride-rich formation waters passing through shaley sedimentary sections could mobilize large amounts of base metals. In discussing carbonate-hosted deposits, we mentioned that such shaley basins are prone to the development of overpressures—pore pressures greater than normal hydrostatic pressure. In the Gulf Coast, such overpressures can build up to the point that they exceed the strength of the overburden, leading to structural disruptions such as faults and mud-cored anticlines (Potter and others 1980, fig. 3.75). This "sedimentary tectonism" can lead to upward migration of large volumes of hot water. A combination of overpressured shales with large faults reaching the surface should be particularly favorable for Pb-Zn mineralization, and most shale hosted Pb-Zn deposits do, indeed, appear to be associated with rifting in an environment of rapid shale deposition (Morganti 1981). Mineralization in rift zones may be additionally favored by higher than normal heat flow.

CHAPTER 8.

Volcanic-Sedimentary Ores

"He got more and more furious as he heard me, so he tore the top
off from a high mountain, and flung it just in front of my ship so
that it was within a little of hitting the end of the rudder. The sea
quaked as the rock fell into it, and the wash of the wave it raised
carried us back towards the mainland, and forced us towards the
shore."

Odysseus deriding Polyphemus in *The Odyssey*, Book IX.
From Samuel Butler (trans.), *The Odyssey of Homer.*
(Roslyn, New York: Walter J. Black, Inc., for the Classics
Club®, 1944, page 115.)
Reprinted with permission of the publisher.

Many ore deposits associated with vocanic rocks show sedimentary features, and
research on modern examples reveals that formation of the ore is partly contemporaneous with deposition of the enclosing sediment. Thus, these ores are partly
syngenetic, but would not be classed as sedimentary by most geologists. Such
deposits easily warrant a separate book; in this chapter I will restrict the discussion
mostly to the sedimentary portions of the deposits and how they compare with
purely sedimentary accumulations. For a long review, one with much attention
to Precambrian examples, which are neglected here, see Franklin and others
(1981). Two other good summaries, which provide more information on the
high-temperature part of these systems, are Klau and Large (1980) and Finlow-
Bates (1980).

Volcanic-sedimentary deposits can be divided into two main groups: those
formed at divergent plate boundaries (the Red Sea, Cyprus), characterized by a
simple suite of metals dominated by Cu, and those formed at consuming plate
boundaries (Kuroko), which contain a variety of metals. The second group can
be subdivided into early island arc deposits, mostly Cu-Zn, and deposits of
mature island arcs, containing Cu-Zn-Ba-Pb. It has become customary to refer
to the mid-ocean ridge deposits as Cyprus-type and the mature island arc deposits
as Kuroko-type, but no accepted model for the intermediate type has emerged.
Many Archean deposits seem to belong to this group, as does the Rambler Mine
in Newfoundland (Table 8-1).

Table 8-1. Three main types and one subvariety of volcanic-sedimentary ore deposits can be distinguished based on composition of the ore and associated volcanics. Characteristics of the two purely sedimentary types are shown for comparison.

| | Depositional Environment | Tectonic Environment | | Volcanic Rocks | Clastic Sedimentary Rocks | Approximate Age Range | Examples |
		General Conditions	Plate Tectonic Setting				
EXHALATIVE VOLCANOGENIC GROUP							
Primitive type Zn-Cu:Ag-Au	Extensive, evolving, deep to shallow water, tholeiitic to calc-alkaline marine volcanism	Major subsidence compression	Subduction at consuming margin, island arc	Fully differentiated suites Basaltic to rhyodacitic	Volcaniclastics Graywackes	Archean — Early proterozoic — Early phanerozoic	Noranda, Kidd Creek — Jerome, Flin Flon Crandon — West Shasta
Polymetallic type Pb-Zn-Cu:Ag-Au	Explosive shallow calc-alkaline-alkaline marine-continental volcanism	Subsidence regional compression but local tension	Back-arc or post-arc spreading; crustal rifting at consuming margin	Bimodal? Suites Tholeiitic basalts Calc-alkalic lavas, pyroclastics	Volcaniclastics Increasing clastics Minimum carbonates	Early proterozoic — Phanerozoic	Prescott, Sudbury Basin — New Brunswick East Shasta Japan
Cupreous pyrite type Cu:Au	Deep tholeiitic marine volcanism	Minor subsidence tension	Oceanic rifting at accreting margin	Ophiolitic suites Tholeiitic basalts	Minor to lacking	Phanerozoic	Newfoundland Cyprus, Turkey, Oman
Kieslager type Cu-Zn:Au	Deep marine sedimentation and tholeiitic volcanism	Major subsidence compression	Fore-arc trough or trench	Mafic; tholeiitic(?) (Amphibolite)	Graywacke, shale(?) (Biot.-amphib. schist)	Late proterozoic — Paleozoic	Matchless-Otjihase S.W. Africa; Ducktown, Tenn; Besshi, Japan; Goldstream, B.C.

Table 8-1. (Continued).

	Depositional Environment	Tectonic Environment		Volcanic Rocks	Clastic Sedimentary Rocks	Approximate Age Range	Examples
		General Conditions	Plate Tectonic Setting				
EXHALATIVE SEDIMENTARY GROUP							
Clastic hosted type Pb-Zn:Ag	Deep marine sedimentation; minor tholeiitic activity	Major subsidence tension, rifting	Separation at continental rift; aulacogenic trough or trench	Minor, basaltic (Gabbroic-amphibolitic intrusive sheets)	Argillite, turbidite	Mid proterozoic	Sullivan, Broken Hill, McArthur River
						Paleozoic	Anvil, Meggen, Rammelsberg, Howards Pass
Carbonate hosted type Zn-Pb:(Ag)	Shallow marine-shelf sedimentation	Minor subsidence, tension	Shelf; local fault-controlled basin	Minor to lacking (tuffaceous beds)	Carbonate lst.-dol. sandstone, shale	Late proterozoic Phanerozoic	Balmat Navan, Silvermines, Tynagh, Ireland

Source: Hutchinson 1980, tables 3 and 4.

All these deposits share a common genetic feature: heated sewater is the transporting fluid. The high heat flow around volcanic centers induces hydrothermal circulation of the water in the surrounding rocks, whether ground water or seawater. Porphyry Cu-Mo deposits are the usual product in terrestrial settings where meteoric water provides the fluid, while Cyprus or Kuroko deposits form in submarine settings with seawater as the fluid (Sillitoe 1980). In the porphyry case, ore deposition is confined to the subsurface; for the submarine deposits, the metals are deposited both within the volcanic mass as stockwork ore and at the surface where the fluids vent into seawater, either as massive sulfides or as Fe-Mn oxides.

I. DEPOSITS OF DIVERGENT PLATE BOUNDARIES

Modern Examples

Beginning in the mid 1960s, geologists became aware of the deposition of metal-rich sediments along mid-ocean ridges (e.g., Bonatti and Joensuu 1966, Boström and Peterson 1966). The prediction that the high heat flow in these areas should induce hydrothermal circulation of seawater through the oceanic crust (Elder 1965, Deffeyes 1970) made it logical to assume that these deposits were related to hydrothermal alteration of basalt by seawater (e.g., Hart 1973, Spooner and Fyfe 1973). These predictions have been dramatically confirmed by the direct sampling and photography of active hydrothermal vents on sites such as the Galapagos Rift (Corliss and others 1979).

The volume of seawater flowing through such systems is large, and may be an important control of seawater chemistry (Maynard 1976, Wolery and Sleep 1976, Edmond and others 1979). The dominant reactions seem to be exchange of Mg^{2+} in seawater for Ca^{2+}, Fe^{2+}, and Mn^{2+} in the basalt. Considerable H_4SiO_4 is also released, and SO_4^{2-} is reduced to H_2S or HS^-, probably by reaction with iron (McDuff and Edmond 1982). In addition to these ridge crest, or axial systems, which operate at around 300°C, there are also off-axis systems that are cooler, about 100-150°C, and produce a release of Mg^{2+} from the basalt rather than an uptake (Donnelly and others 1979, 1980). The balance between these two kinds of system probably results in little net exchange of Na^+, K^+, Ca^{2+}, or Mg^{2+} between oceanic crust and seawater (Wolery and Sleep in press), but there should still be an appreciable net addition of Fe^{2+}, Mn^{2+}, Ba^{2+}, and H_4SiO_4 to seawater from the axial vents. The insolubility of the Fe, Mn, and Ba in seawater, under oxidizing conditions, leads to their precipitation and consequent enrichment in ridge-crest sediment. In addition, the reduction of SO_4^{2-} to sulfides in the high-temperature systems can lead to the precipitation of Cu or Zn, either within the hydrothermal passageways or near their exits. The off-axis systems apparently do not produce mineralization.

As discussed in Chapter 5, Fe^{2+} oxidizes at a lower Eh than Mn^{2+}, so that

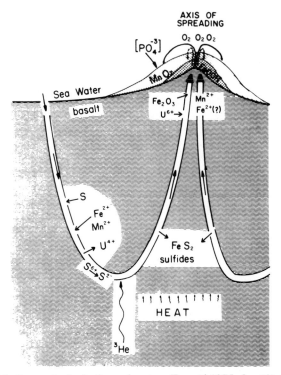

Fig. 8-1. Model of seawater-hydrothermal system (Bonatti 1975, fig. 19). The horizontal and vertical scales have not yet been determined, but might be 100–200 m across the top by 2 km deep. Note the separation of Fe and Mn.

Fe precipitates first in an Eh gradient, and Mn tends to be dispersed farther from the vents (Fig. 8-1). The available data for Ba (Table 8-2) suggest that it follows Mn. Thus, there is a facies sequence of Fe → Mn + Ba that can be useful in exploration, and is one indicator of a volcanic source for an orebody.

So far, the only metal sulfides found in these deposits have been Cu and Zn (Bonatti and others 1976a, Francheteau and others 1979, Hékinian and others 1980). The most abundant minerals are chalcopyrite, sphalerite, wurzite, and pyrite accompanied by marcasite and barite. Anhydrite is common around active vents, but appears to dissolve rapidly once the vent stops (Haymon and Kastner 1981); other alterations are the formation of atacamite, tenorite, jarosite, and Cu-rich iron hydroxides. Samples from the Mid-Atlantic Ridge contain only Cu-Fe sulfides (Bonatti and others 1976a), whereas the sulfides that have been observed forming on the East Pacific Rise are predominantly Zn and Fe, with subordinate Cu (Hékinian and others 1980). A different level may have been sampled at the two sites, with Cu predominating at depth in the Pacific deposits. Alternatively, there may be a real difference in chemistry as a result of the higher spreading rate in the Pacific, which also affects the Fe-Mn sediments (Fig. 8-2).

Table 8-2. Selected analyses of metalliferous sediment from modern mid-ocean ridge deposits. (A blank indicates not sought; a zero, a trace or none detected.)

| | Galapagos | | East Pacific Rise | | Gulf of Aden | |
| | todorokite-rich | smectite-rich | "gossan" | clay | oxide | clay |
Major Elements (%)						
Na	3.0	1.1	0	1.0	3.0	1.2
Mg	1.4	1.5	0.4	0.3	1.8	1.8
Al	0.2	0.1	0.3	0.3	0.7	0.2
Si	0.8	22.0	5.3	18.6	7.3	22.0
K	0.6	1.5	0.2	0.9	1.4	3.2
Ca	1.5	0.5	0.3	0.7	1.5	0.5
Mn	50.0	0.2	0	0	32.9	0
Fe	0.3	25.4	39.6	19.8	2.7	22.2
S			1.8	0.1	0.1	0.05
Trace Elements (ppm)						
Co	5	1			30	5
Ni	470	15			400	4
Cu	100	4	0			0
Zn	380	37		50,000	310	0
As	14	7				
Sb	26	1				
Ba	1700	250			1180	70

Source: Galapagos = Corliss and others 1978, table 1; East Pacific Rise = Hekinian and others 1980, table 5; Gulf of Aden = Cann and others 1977, tables 1 and 2.

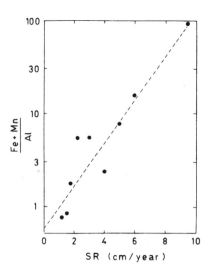

Fig. 8-2.
Higher spreading rates (SR) result in more intense mineralization, as shown by the proportions of hydrothermal (Fe + Mn) and detrital (Al) sediment on mid-ocean ridges (Bostrom 1973, fig. 2). Reproduced, with permission, from the Annual Review of Earth and Planetary Sciences, Volume 3. © 1975 by Annual Reviews, Inc.

In the East Pacific Rise deposits, pyrite and marcasite make up about half of the sulfides, sphalerite and wurzite about 34 percent, and Cu minerals the rest (Hékinian and others 1980). The Cu occurs as Cu-rich pyrite, with Cu contents between 2 and 10 percent, and as chalcopyrite. Trace amounts of digenite and bornite also are found. The sulfides are interlaminated with Fe hydroxides in distinctive tubes that range in size from 0.5 to 3 mm across through which the hydrothermal fluids flowed. In addition to these tubes, there are several other biogenic and abiogenic structures observed around the vents (Fig. 8-3). Considerable deterioration of these features occurs after hydrothermal activity ceases, so it is not yet known which are likely to be preserved in ancient deposits.

Because sulfur isotopes are uniformly more positive than those of S in basalt (Fig. 8-4), seawater sulfate must be a major source of sulfur, but the lack of data on the oxygen fugacity and pH of the systems at depth makes it impossible to quantify the relative contributions of S from the basalt and from seawater (Styrt and others 1981). Further, the lack of fractionation between pyrite and sphalerite suggests that they were not deposited in isotopic equilibrium, nor does there seem to be equilibrium between sulfides and sulfate (Sakai and others 1980). It is also not known whether the unusual biological activity found with the vents has any effect on mineral precipitation.

Experimental work on basalt-seawater reactions sheds further light on how these deposits form. For the mobilization of heavy metals from basalt, two factors seem to be important: temperature and the water/rock ratio (Mottl and Holland 1978, Mottl and others 1979, Seyfried and Bischoff 1977, 1979, 1981). Either high temperatures ($>400°C$) or high water/rock ratios (>10 at $300°C$) are required. At the higher temperatures, complexing of the metals by the chloride in seawater becomes significant, and a large proportion of the metals in the basalt can be leached. At lower temperatures, however, the chloride content of seawater is too small for significant complexing, and no leaching takes place unless the pH is low. Lowering of pH in experimental systems is due to reactions like

$$3Mg^{2+} + 4H_4SiO_4 \rightarrow Mg_3Si_4O_{10}(OH)_2 + 4H_2O + 6H^+,$$

reaction of Mg^{2+} from seawater with silica released from hydration of the basalt to make a secondary phase. Here, I have used talc as a model compound, although the actual products are more complex substances such as smectites. In a rock-dominated system, the Mg^{2+} from seawater is quickly absorbed, no net acid production occurs, and metals are not solubilized, but in a system with a high water/rock ratio, sufficient seawater Mg^{2+} goes through the system to keep the pH low—5 or less—and, therefore, to solubilize appreciable quantities of metals (Bischoff and Dickson 1975; Seyfried and Bischoff 1977). Table 8-3 shows that, indeed, NaCl-basalt systems do not become acid, and no metals are brought into solution unless Mg^{2+} is present. Because most of these systems are thought to operate at 300-350°C, chloride complexing is probably not important in leaching and transporting the metals. Also, the low pH indicates that the metals and reduced S could be transported in the same fluid (Anderson 1975).

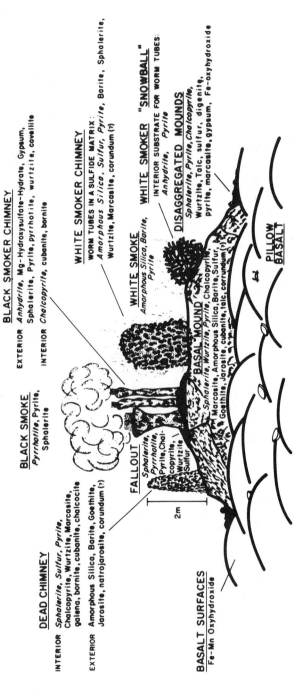

Fig. 8-3. Types of hydrothermal structures observed on East Pacific Rise. These mounds may be analogous to the Cyprus "ochres" (Haymon and Kastner 1981, fig. 10).

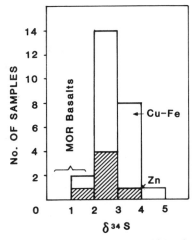

Fig. 8-4. Sulfur isotopic composition of Fe-Cu-Zn sulfides from the East Pacific Rise. (Data of Styrt and others 1981, fig. 5.) Hekinian and others (1980, p. 1442) cite the following averages: pyr = +3.45, sph = +2.50, cpy = +2.1. The positive values relative to mid-ocean ridge basalts indicate that at least some S comes from seawater.

Divergent Boundaries: Ancient Examples

There are several ancient analogues of these deposits, the best known being the ophiolite-associated copper orebodies of Cyprus (Hutchinson and Searle 1970, Constantinou and Govett 1973). Others are known from Newfoundland (Upadhyay and Strong 1973, Duke and Hutchinson 1974) and from the Apennines (Bonatti and others 1976b). In Oman, similar deposits have been found (Fleet and Robertson 1980, Pearce and others 1981), but they seem to have been deposited during a succeeding period of island-arc formation, rather than during the spreading event with which the ophiolites are associated. Ophiolites, which are interpreted to be former oceanic ridges now exposed on land, have a characteristic stratigraphy (Fig. 8-5). Metal enrichment occurs as epigenetic and syngenetic pyrite-chalcopyrite at the top of the sheeted dike complex and in the basal pillow lavas, and as syngenetic Fe-rich or Mn-rich accumulations either directly overlying the sulfides or higher in the pillow-lava sequence (Fig. 8-6). In all cases studied so far, the ores occupy this stratigraphic position, which is a valuable aid in exploration (Strong 1973). In contrast to some of the modern examples, Cu greatly predominates over Zn in all cases.

The oxide portions of the Cyprus deposits have been described by Constantinou and Govett (1972), Elderfield and others (1972), Robertson and Hudson (1973) and Robertson (1975). Two forms occur: Fe-rich deposits (ochres) directly overlying the massive sulfides and containing some intermixed sulfides, and Mn-rich deposits (umbers) higher in the section, within or at the top of the Upper Pillow Lavas (Fig. 8-7). The ochres are thought to be products of submarine alteration

Table 8-3. Results of reaction between hot seawater and basaltic glass at high water/rock ratios (T = 300°C, ratio = 10/1 by weight).

Time	pH	Composition (ppt)									
		Na	K	Mg	Ca	SO_4	SiO_2	H_2S	Fe	Mn	Ba
Seawater											
Initial	7.72	10.3	0.38	1.24	0.39	2.60	0	0	0	0	0.01
24 hr	4.83	10.8	0.41	0.07	0.96	0.08	1.72	0.13	35.5	17.0	0.5
2200 hr	5.62	11.0	0.49	0	0.82	0.02	0.58	10.4	3.5	0.05	0.5
0.45M NaCl											
Initial	6.00	10.1	0	0	0	—	0	0	0	0	0
36 hr	6.98	10.03	0.12	0	0.64	—	0.50	0.02	0	0	0
564 hr	6.31	0.96	0.15	0	0.42	—	0.57	0.01	0	0	0

Source: Seyfried and Bischoff 1981, tables 1 and 3.

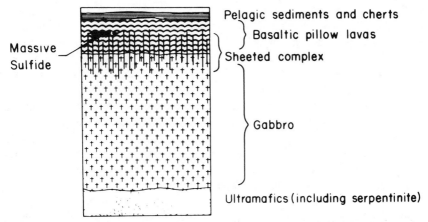

Fig. 8-5. Generalized stratigraphy of an ophiolite, with location of pyrite-chalcopyrite orebodies. Total thickness of the section is about 8 km (Sawkins 1976a, fig. 2).

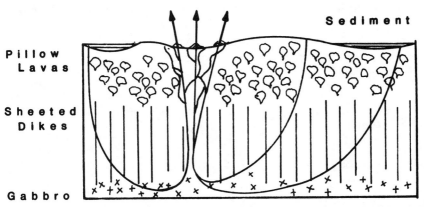

Fig. 8-6. Hydrothermal circulation pattern inferred for the Troodos Complex (from Spooner and Bray 1977, fig. 1).

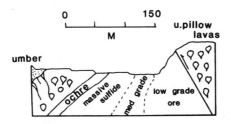

Fig. 8-7. Relationship of umber and ochre to sulfide mineralization in a typical Cyprus Cu mine (after Robertson 1975, fig. 5c).

of the massive sulfides, and are similar in many respects to the hydrothermal oxide deposits of the East Pacific Rise (Hékinian and others 1980). The umbers, in contrast, are richer in Mn, which imparts a chocolate-brown color, and are interbedded with pelagic sediment. Chemically, they are similar to the ochres, but with a higher Mn/Fe ratio (Table 8-4). Note, also, the higher Ba in the umbers, although still low compared with deep-sea clay, and the enrichment of both umbers and ochres in V, Cu, and Zn, but not Pb. In one locality, tubular structures of from 0.5 to 1 cm in diameter overlie brecciated pillow lavas (Robertson 1975, p. 519). They may be analogues of similar features described from the East Pacific Rise (Hékinian and others 1980); if so, they support a hydrothermal origin for the umbers, as does the presence of brecciation and veining of the lava beneath some umber deposits (Fig. 8-8). Because the umbers are found higher in the stratigraphic sequence, separated from the ores by the Upper Pillow Lavas, they probably represent a separate event in time, but one related genetically to ore deposition. Robertson (1975, p. 528) proposed that they formed from hydrothermal solutions released into oxidizing seawater on the elevated flanks of the ridge, whereas the massive sulfides formed from solutions released into small anoxic basins within the axial rift. Subsequently, these became superficially oxidized to ochres.

Fluid inclusions, S isotopes, and Sr isotopes indicate that seawater was the hydrothermal fluid. Spooner and Bray (1977) showed that fluid inclusions in

Table 8-4. Partial chemical analyses of Cyprus umbers and ochres.

Major Elements (%)	Umbers	Ochres	Average Deep-Sea Clay
Mg	0.7	0.5	2.1
Al	2.3	3.6	8.5
Si	9.8	—	25.0
K	0.7	0.6	2.5
Ca	1.1	0.8	2.9
Mn	5.8	0.8	0.7
Fe	27.9	47.6	6.5
Mn/Fe	0.21	0.02	0.11
Minor Elements (ppm)			
V	760	870	120
Cr	30	40	90
Co	100	50	70
Ni	210	130	220
Cu	900	1600	250
Zn	240	590	170
Sr	410	60	180
Zr	370	270	150
Ba	430	20	2300
Pb	160	60	80

Source: Robertson 1978, table 2.

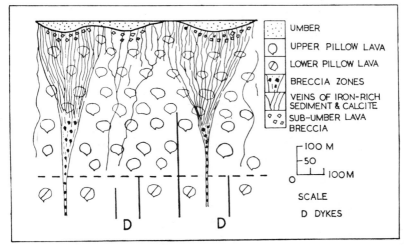

UMBER

UPPER PILLOW LAVA

LOWER PILLOW LAVA

BRECCIA ZONES

VEINS OF IRON-RICH
SEDIMENT & CALCITE

SUB-UMBER LAVA
BRECCIA

100 M
50
100M
0
SCALE

D DYKES

Fig. 8-8. The Cyprus umbers commonly show features such as brecciation and veining of the underlying pillow lavas that suggest a syngenetic-hydrothermal origin, related to that of the massive sulfides (Robertson 1975, fig. 8).

quartz co-precipitated with the ore have a salinity indistinguishable from that of seawater. As in the modern examples, S isotopes indicate that S from seawater was incorporated into the sulfides. If seawater SO_4^{2-} is reduced to sulfide by the iron in basalt at high temperatures, there is an isotope fractionation that depends on the pH and oxidation state of the system (Fig. 8-9). For the ophiolitic Cu deposits of Notre Dame Bay, Newfoundland, Bachinski (1977) has argued, based on the occurrence of muscovite rather than kaolinite or K-feldspar as an alteration product, that the pH in the hydrothermal fluids was between 4.5 and 6 at 250°C, a value consistent with the experimental results of Seyfried and Bischoff (1981). Similarly, the presence of pyrite as the major iron-bearing species sets limits on the fugacity of oxygen (Fig. 8-9). Under these conditions, tholeitic basalt should yield S between -25.5 and -4 permil, whereas S from Ordovician seawater ($\delta^{34}S = +27$), should fall in the range $+1.5$ to $+23$. The observed values suggest (Table 8-5) that most, if not all, the S came from seawater. For Cyprus, the same conclusion can be drawn, allowing for lighter seawater SO_4^{2-} in the Cretaceous ($+15$ permil). $^{87}Sr/^{86}Sr$ ratios, which in some samples are close to that of Cretaceous seawater, also indicate a large seawater component with water/rock ratios exceeding 15/1 (Chapman and Spooner 1977, Spooner and others 1977).

Divergent Boundaries: The Red Sea

Perhaps the best-studied deposits associated with oceanic ridges are those in the Red Sea, yet they are somewhat anomalous because of the proximity of continental land masses, and the apparent involvement of this continental material in their genesis. Good descriptions of the deposits and their inferred origin can be found

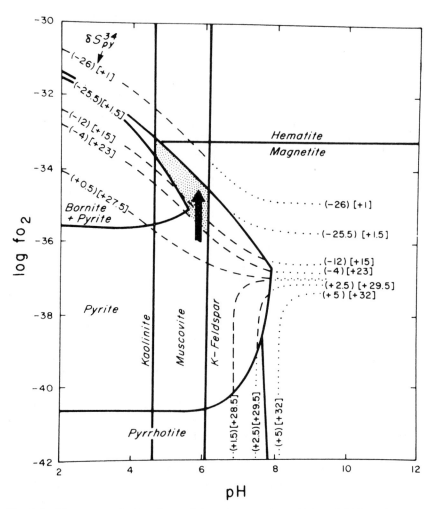

Fig. 8-9. S isotopes and mineral assemblages in a massive sulfide at 250°C, as functions of pH and oxygenation. The round brackets indicate δ³⁴S of pyrite precipitated from a solution with S at +27 permil (Ordovician seawater); square brackets with S at 0 permil (basalt). The arrow shows inferred direction of change during development of the orebody; shaded area range of conditions (Bachinski 1977, fig. 8).

in Bischoff (1969), Hackett and Bischoff (1973), and Shanks and Bischoff (1977). Overlying the metal-rich sediments is a dense hot brine, from which they are believed to have been deposited. This lower brine is, in turn, separated from seawater by a second brine of intermediate composition, which is probably a mixture of the bottom brine and seawater (Craig 1969, Brewer and Spencer 1969). Chemical analyses (Table 8-6) show the depletion in Mg^{2+} and enrichment in Ca^{2+}, compared with seawater, that is typical of basalt-seawater interactions, but Na^+ and Cl^- are unusually high. This enrichment is believed to be caused

Table 8-5. Isotopic composition of sulfur from ophiolitic copper deposits.

Notre Dame Bay, Newfoundland	
(pyrite and chalcopyrite)	
Tilt Cove	+ 5.5 to +23.0
Lush's Bight	
Rendell Jackman	+ 9.1 to +18.0
Little Bay	+11.4 to +13.8
Whalesback	+ 3.6 to + 7.0
Average of all samples	+ 9.0
Contemporaneous seawater	+27.0
Cyprus	
pyritic ironstones	+ 1.9 to + 5.0
massive pyrite	+ 3.0 to + 5.3
stockwork pyrite	+ 4.9 to + 7.0
Average of all samples	+ 4.8
Contemporaneous seawater	+15.0

Source: Bachinski 1977, tables 5, 7.

by dissolution of Miocene evaporites bordering the Red Sea. Note also the greater dispersion of Mn than Fe into the upper brine, and the depletion in SO_4^{2-}, which may be caused by deposition of anhydrite (Shanks and Bischoff 1980, p. 456).

The sediment precipitated from these brines has been divided into seven facies (Fig. 8-10), but three predominate (Table 8-7): Fe-smectite, amorphous/goethite, and sulfide. The last two are found in Cyprus-type deposits; all three are reported

Table 8-6. Composition, in ppm, of Red Sea brines and seawater.

	Atlantis II 56°C Brine	Atlantis II 44°C Brine	Red Sea Deep Water
Na	92,600	46,900	12,500
Mg	764	1,190	1,490
Si	28	—	3
K	1,870	1,070	450
Ca	5,150	2,470	470
Cl	156,030	80,040	22,500
SO$_4$	840	2,260	3,140
HCO$_3$	140	160	140
pH	5.61	6.50	8.17
Fe	81.0	0.2	0.01
Mn	82.0	82.0	0.005
Cu	0.26	0.02	0.01
Zn	5.4	0.15	0.02
Ba	0.9	—	0.01
Pb	0.63	0.009	0.00003

Source: Shanks and Bischoff 1977, table 1; Hartmann 1969, table 2.

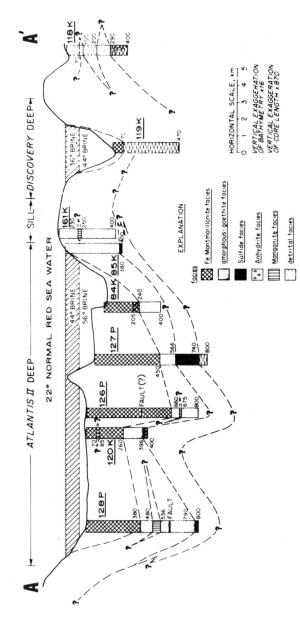

Fig. 8-10. Distribution of facies of metalliferous sediment in the Red Sea. These facies were deposited at different times under slightly different conditions rather than being lateral equivalents of one another (Bischoff 1969, fig. 2).

Table 8-7. Average chemical analyses (in %) of Red Sea metalliferous sediment.

	Detrital	Fe-Smectite	Goethite/amorphous	Sulfide
SiO_2	27.3	24.4	8.7	24.7
Al_2O_3	8.4	1.7	1.1	1.5
Fe_2O_3	4.9	24.1	61.2	9.4
FeO	1.4	11.7	2.7	13.4
MnO_2	0.7	2.4	1.3	1.3
CaO	23.6	4.8	3.4	2.5
CO_2	23.1	8.6	3.6	5.7
S	0.3	3.9	0.6	16.8
Zn	0.06	2.6	0.6	9.8
Cu	—	0.6	0.2	3.6
Pb	0.03	0.1	0.1	0.2

Source: Bischoff 1969, table 3.

from the modern ridge-crest deposits (Hékinian and others 1980, Oudin and others 1981). As in the other deposits, the Red Sea sediments are rich in Fe relative to Mn, which, instead, is dispersed into the surrounding sediments, forming a halo of Mn-enrichment around the ores (Bignell and others 1976, Cronan 1976, 1980).

By assuming that the lower brine in the Atlantis II deep is a cooled sample of the hydrothermal fluid, Shanks and Bischoff (1977) have reconstructed the conditions of metal transport and deposition in this system. The results confirm the idea (Craig 1966, 1969, Manheim 1974) that the brines were derived from normal seawater that acquired a high salinity through reaction with evaporites at low temperatures, then was heated to about 200°C by interaction with recent intrusive rocks of the rift zone at high water/rock ratios. In their model of the fluid, S is present mostly as sulfate complexes such as $NaSO_4^-$ and $MgSO_4^0$; H_2S is only 2 ppm. Heavy metals occur mostly as chloride complexes at higher temperatures, but cooling below 150°C leads to dissociation and precipitation of the metals as sulfides. Note that the excess of metals over reduced S should lead to precipitation of virtually all of the S, and considerable loss of metals by dispersion into surrounding sediments and seawater. The calculated sequence of precipitation (Fig. 8-11) implies that anhydrite and chalcopyrite should be largely confined to the subsurface portion of the deposit, with the sphalerite and galena precipitating mostly after the hydrothermal fluid exits onto the sea floor. The high density of the brine layer restricts circulation, so that the bottom water becomes anoxic and the sulfides are protected from oxidation. Fig. 8-12 shows undersaturation with respect to siderite and barite at all temperatures, yet both are common. The siderite in the Red Sea deposits is a Mn-rich variety that might have a lower solubility than pure Fe siderite; the barite probably forms at the interface between the two brines, incorporating SO_4^{2-} from seawater. Unlike the barite at Tynagh (Chapter 7), S in barite from the Red Sea is lighter than seawater by 5 to 15 permil (Shanks and Bischoff 1977, fig. 8) suggesting that some of the S must have come from oxidation of sulfides at the brine interface.

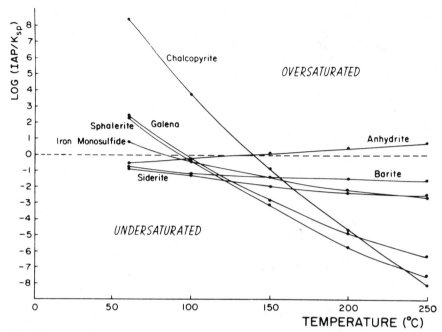

Fig. 8-11. Degree of saturation of the Red Sea brines with respect to various minerals as a function of temperature. IAP/K is the ratio of the measured to the equilibrium activities of the elements that make up the minerals. The brine is thought to have been heated to at most 200–250°C and vented at 50–100°C. Thus the Fe, Zn, and Pb sulfides would mostly be precipitated after venting onto the seafloor, while Cu would be confined to the subsurface (Shanks and Bischoff 1977, fig. 7).

The sulfides themselves have S confined mostly to $+1$ to $+15$ permil, similar to other seawater-basalt hydrothermal systems (Shanks and Bischoff 1980, fig. 6). SO_4^{2-} in the brine is $+20.3$ permil, indistinguishable from overlying Red Sea water, suggesting that only a small fraction of the SO_4^{2-} in the hydrothermal system is reduced by reaction with the basalt, and thus that the f_{O_2} of the system is relatively high. If the source SO_4^{2-} for the sulfide minerals does have $\delta^{34}S$ of $+20.3$ permil, then the sulfides are somewhat heavy compared with expected fractionation between sulfate and sulfide at 250°C. Apparently, exchange between $NaSO_4^-$ and H_2S is slow at these temperatures, and a kinetic isotope effect takes over (Shanks and Bischoff 1980, p. 456). In addition to the hydrothermal sulfides, some facies of the deposits have pyrite with very light S (-35 to -25), indicative of bacterial sulfate reduction. Thus, like Tynagh and McArthur River (Chapter 7), there is evidence of two sources of S.

Similar deposits are known from land areas around the Red Sea or in related rift systems. Two of the deeper lakes in the East African rift valley show evidence of hydrothermal mineralization: Lake Kivu (Degens and others 1972, 1973, Degens and Kulbicki 1973, Degens and Ross 1976), and Lake Malawi (Müller and Förstner 1973). Because of stagnant bottom water, these lakes can accumulate reduced minerals such as pyrite. Other common phases are amorphous silica, nontronite, limonite, and vivianite. Bordering the Red Sea in the Afar Rift is a small deposit of probable volcanic-sedimentary origin that may be an exposed version of those in the Red Sea (Bonatti and others 1972a). Here, about 2 m of Mn and Fe oxide overlie altered basalt, and are succeeded by normal marine sediments. The Fe-rich layers consist of goethite and an expandable Fe^{3+} clay, probably nontronite. The normal and expanded basal spacings are 14.8 and 16.4Å, and the spacing of the (060) reflections suggest a dioctahedral site occupancy. In contrast, nontronite from the Red Sea and Lake Malawi have some ferrous iron (Table 8-8), and the basal spacings for the Red Sea clays are smaller, 13.5 and 15.5Å. Also, the (060) reflections indicate a partial trioctahedral character. In the Mn-rich layers, well-crystallized pyrolusite predominates, except in occasional porous zones which contain poorly crystalline birnessite, todorokite and pyrolusite. Strontium-rich barite is a common accessory in the Mn-rich layers, and Ba correlates with Mn/Fe ratio (Fig. 8-12).

From the above, it can be seen that deposits associated with accreting plate boundaries can be wholly oceanic (the East Pacific Rise deposits, Cyprus) or have appreciable continental influence (Red Sea, East African Rift). The former probably involve hotter fluids (300-350°C) of close to seawater salinity; the hydrothermal fluids in the latter case are somewhat cooler (<250°C), and of a wide range of salinities, from hypersaline to fresh water. With the exception of the deposits in the East African Rift lakes, all show an intimate association with basaltic volcanism. Further, their mineralogy is similar: Fe-smectite, Fe, Cu, and Zn sulfides, and Fe and Mn oxides are found in almost all cases. Normally, Fe and Cu greatly predominate over Zn in the oceanic deposits, but the deposits on the East Pacific Rise are a conspicuous exception that needs more study.

Table 8-8. Chemical analyses of nontronites and limonite from Lake Malawi and the Red Sea.

	Malawi Nontronite	Red Sea Nontronite	Malawi Limonite
SiO_2	42.1	29.3	20.4
Al_2O_3	4.7	1.9	3.2
Fe_2O_3	29.0	32.9	50.5
FeO	0.5	7.0	0.21
MnO	0.15	0.35	0.31
MgO	0.85	0.81	0.31
CaO	1.15	1.8	0.84
Na_2O	0.45	0.65	0.23
K_2O	0.58	0.5	0.31
TiO_2	0.26	—	0.15
P_2O_5	0.52	—	0.82
H_2O^+	19.8	19.7	18.3
Cu	—	0.81	—
Zn	—	4.4	—

Source: Müller and Förstner 1973, table 3.

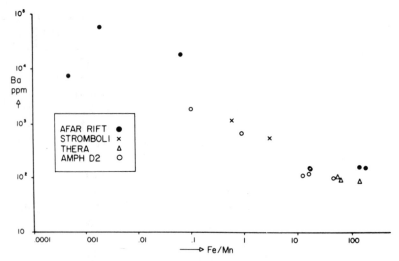

Fig. 8-12. Barium increases as the proportion of Mn in an Fe-Mn deposit increases. Because the Mn is dispersed farther from hydrothermal vents than Fe, Ba also should show a halo of enrichment around volcanic-sedimentary ore deposits (Bonatti and others 1972b, fig. 6). © 1972 by Dowden, Hutchinson & Ross, Inc., Stroudsburg, PA. Reprinted by permission of the publisher.

One additional group of deposits associated with basic volcanism needs mention: the iron ores of Central Europe of the Lahn-Dill type. The tectonic setting in this case is not so clear-cut because the volcanics overlie a thick cratonic sequence, but do not occupy a narrow rift zone. Probably the subsidence and volcanism in this area was caused by back-arc spreading related to a subduction zone to the south (see Anderton and others 1979, p. 173 for an illustration of this geometry). The ores are Middle Devonian to Early Carboniferous in age (Kräutner 1977, fig. 14), and contain no metal enrichments other than iron; the massive sulfide and Mn-rich facies seen in most other volcanic-sedimentary deposits are not developed. Chemically, the ores are highly variable, but are usually rich in either silica or calcium carbonate (Table 8-9). These two substances

Table 8-9. Chemical analyses of some of the ore types of the Lahn area.

	Silica-Rich Hematite	Resedimented Hematite	Sideritic
SiO_2	53.2	21.9	3.8
Al_2O_3	—	—	1.0
Fe_2O_3	42.7	50.7	1.6
FeO	1.5	1.8	43.5
MgO	—	—	—
CaO	0.2	7.7	9.7
Mn	0.09	0.12	0.16
P	0.06	0.10	0.19
S	—	—	—

Source: Quade 1976, table 2.

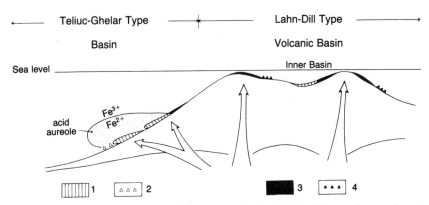

Fig. 8-13. Model for formation of iron ores at Lahn-Dill. 1. Primary iron carbonate. 2. Resedimented iron carbonate. 3. Primary iron oxide. 4. Resedimented iron oxide (Kräutner 1977, fig. 13).

are negatively correlated (Lutzens and Burchardt 1972, table 3), with a tendency for the ores to become less siliceous and more calcareous upwards in vertical profile. Hematite predominates, accompanied by magnetite, siderite, and chamosite. Chamosite, where present, is found at the ore-volcanics contact; siderite appears to have formed where the hydrothermal fluids vented into deeper water (Fig. 8-13), while the oxide ores were localized on the crests of volcanic rises, where they were commonly succeeded by reefs. It is not known why this setting should have produced only iron enrichment.

II. DEPOSITS OF CONVERGENT PLATE BOUNDARIES

Modern Examples

There are some descriptions of deposition of metalliferous sediment in association with volcanoes of island arcs, but, so far, there has been none of the direct underwater observation so helpful in understanding the oceanic-ridge deposits. Two areas have been investigated: the Mediterranean and the island of New Britain in the Bismarck Archipelago. For the most part, the deposits consist only of Fe-Mn oxides, with minor sulfides and carbonates; Fe-smectites have not yet been positively identified. In contrast to the divergent-margin rocks, the volcanics in this case are andesites or dacites.

Metalliferous sediment around the Mediterranean islands of Stromboli and Santorini have been described by Bonatti and others (1972b), Puchelt (1973), and Puchelt and others (1973). At Stromboli, the metalliferous sediment is mostly Fe-Mn oxide (Table 8-10). X-ray diffraction shows only birnessite, but the presence of cryptocrystalline goethite and amorphous SiO_2 is inferred. On Santorini, the metalliferous sediment is up to 3 m thick, has a high water content, and is X-ray amorphous. Little Mn is present (Table 8-10), and CO_2 is much more abundant than at Stromboli, indicating the presence of $FeCO_3$. Scanning

Table 8-10. Chemical analyses of major and minor elements from metalliferous sediments of the Mediterranean islands of Stromboli and Santorini.

	Stromboli	Stromboli	Santorini[1]
Major elements (%)			
Na	1.7	2.2	0.3
Mg	0.4	0.7	0.9
Al	0.4	0.2	0.6
Si	7.9	5.1	13.0
K	0.5	0.8	0.2
Ca	0.4	0.5	0.9
Mn	9.2	22.4	0.2
Fe	28.0	14.3	26.6
P	0.4	0.3	0.1
CO_2	—	—	12.2
H_2O	8.6	—	4.3
Mn/Fe	0.33	1.6	0.001
Trace elements (ppm)			
Ti	60	60	2500
V	90	55	70
Cr	<5	<5	7
Co	15	30	20
Ni	10	15	—
Cu	160	420	20
Zn	—	—	80
Zr	40	70	—
Mo	—	—	40
Ba	550	1070	—

Source: Bonatti and others 1972b, table 1; Puchelt and others 1973, tables 1, 2.
[1]Corrected for seawater using Cl values.

electron microscope studies (Puchelt and others 1973) reveal the presence of small spherical masses, 10 to 20 micrometers across, of siderite crystals, with crystallites about 1 micrometer across. Other phases present are opal, ferrous hydroxide, and vivianite.

Sulfides are conspicuously rare at these two localities, but are common in sediments bordering the island of Vulcano, near Stromboli (Honnorez and others 1973). Abundant fumaroles and hot springs occur on the island and in the surrounding water; the submarine activity is confined to depths of less than 15 m in an elongate zone about 100 m wide, parallel to the shoreline. Chemically, the fumarolic gases are mostly CO_2 and H_2O with about 0.04 percent H_2S (Honnorez and others 1973, table 1). Temperatures range from 100 to 600°C, and mixing of the fumarolic gases with seawater lowers its pH to as low as 2. In the sediment, the result is a sulfidation of Fe-Ti grains, cementation of quartz sand grains by pyrite-marcasite, and silicification of volcanic rock fragments. No base-metal sulfides have been reported. Mn is slightly enriched in the surrounding sediments, and reaches its highest amount, 1900 ppm, in the deepest, most distal part of the bay in which the fumaroles occur (Valette 1973, fig. 3), a distribution consistent with the Mn dispersion seen in other deposits. Ba, however, does not follow Mn, but is highest (1000 ppm) near the shoreline around the fumaroles (Valette 1973, fig. 7).

On New Britain, metalliferous sediment has been reported from two localities. At the first, Matupi Harbor, hot, acid springs and fumaroles are precipitating Fe-Mn oxides in the shallow-marine environment, but no sulfides (Ferguson and Lambert 1972). Based on Cl content (Table 8-11), the geothermal fluid seems to be mostly seawater, modified by acquisition of heavy metals. Note that the strong depletion in Mg^{2+} seen in the basalt cases does not seem to occur here, although Ca^{2+} is still enriched. Typical Mg^{2+} depletion—Ca^{2+} enrichment does occur, however, in seawater thermal springs along the Japanese coast (Sakai and Matsubaya 1974), so the New Britain situation may be exceptional. Sediments precipitated from these springs are rich in Fe relative to Mn, and sometimes contain abundant Zn (Table 8-12). At Talasea (Ferguson and others 1974), Fe sulfides are forming in addition to the oxides. They make up as much as 10 percent of the sediment, but occur as discrete grains, often framboidal, rather than as cements and replacements like those at Vulcano. Both pyrite and marcasite are found, as well as precursor FeS.

None of these modern occurrences show the complex sulfide-barite-anhydrite-Fe oxide assemblages found in ancient deposits from this setting. Therefore, it is uncertain to what extent the features seen in the modern deposits can be used to interpret the ancient ones. Clearly, this is an area in which much fruitful work can be done.

Table 8-11. Chemistry of hot springs discharging at the shoreline in Matupi Harbor, New Britain[1] and at Talasea.[2]

	Tavurvur	Rabalankaia	Talasea	Seawater
Na	13,600	8,600	3,200	10,600
Mg	1,340	1,370	357	1,350
K	756	525	170	380
Ca	395	1,030	365	400
Mn	111	2.7	1.50	0.002
Fe	97	3.5	0.04	0.01
Cu	0.05	0.05	0.01	0.003
Zn	2.53	0.03	0.05	<0.01
As	0.02	0.01	—	0.003
Pb	0.09	0.07	0.02	0.00003
SO_4^{2-}	5,420	2,230	475	2,649
Cl^-	22,500	18,000	6,290	19,000
Na/Cl	0.604	0.478	0.509	0.558
Mg/Cl	0.060	0.076	0.057	0.071
K/Cl	0.034	0.029	0.027	0.020

Sources: 1. Ferguson and Lambert 1972, table 2.
        2. Ferguson and others 1974, table 1.

Table 8-12. Metals found in hot spring precipitates in Matupi Harbor, New Britain.

	Tavruvur	Rabalankaia	Harbor Sediment	Continental Crust
Fe (%)	25.6	17.6	7.95	5.6
Mn (%)	1.03	0.068	0.48	0.09
Zn (ppm)	1150	34	300	70
Cu (ppm)	26	18	50	55
Pb (ppm)	<100	<100	<100	13
Mn/Fe	0.040	0.004	0.060	0.016

Source: Ferguson and Lambert 1972, tables 4, 5.

Convergent Boundaries: Ancient Examples

Subduction-zone volcanics have long been known as favorable hosts for a variety of ore deposits. Of greatest interest are porphyry coppers, mostly found in continental margin volcanic arcs such as the Andes, and massive sulfide deposits of the Kuroko type, mostly found in island arcs. Sillitoe (1980) has suggested that porphyry coppers develop beneath stratovolcanoes, but Kuroko-type deposits form only in resurgent calderas, and that these develop during tensional events in the evolution of the island arc. This hypothesis also explains the association of porphyry coppers with andesites and dacites, of Kuroko ores with rhyodacites and rhyolites.

Earlier, we alluded to a division of massive sulfide deposits of island arcs into those found in "primitive" arcs, containing only Cu and Zn, and those in "mature" or "evolved" arcs, containing commercial concentrations of Pb and Ba as well. (For a discussion of chemical differences in the volcanic rocks of these two classes, see Donnelly and Rogers 1978.) Almost all Archean deposits fall into the first group (Sawkins 1976a, table 1), suggesting that the presence or absence of continental crust is the determining factor. In mature island arcs like Japan, which include appreciable continental crust, deposits of complex chemistry form. Conversely, the small area of continental crust in the Archean should have resulted in few subduction zones of the mature type. Probably the deposits of the "Besshi" type, as defined by Sawkins (1976a, table 3) also belong to the primitive group, but are distinguished by an association with thick graywacke sequences. So far as is known, the ore deposits in both the primitive and mature settings are nearly the same, except for the more complex mineralogy in the mature case. Therefore, because they are much better described, we will only discuss deposits of the Kuroko type in this section. For a description of the setting of ores in the Rambler Mines of Newfoundland, which are in primitive island-arc volcanics, see Tuach and Kennedy (1978); for an excellent discussion of the evolution of volcanic-sedimentary ores of Newfoundland, which display the whole progression from ophiolitic to early to late island arc, see Strong (1977) and Kean and others (1981).

Table 8-13. Facies of mineralization in Kuroko orebodies, from top.

Fe-chert	Cryptocrystalline quartz with lesser amounts of hematite, and sometimes manganite or braunite.
Barite ore	Thin, well-stratified, nearly pure barite.
Black ore (Kuroko)	Stratiform, containing barite, sphalerite, galena, pyrite, and chalcopyrite.
Yellow ore (Oko)	Stratiform, containing mostly pyrite and chalcopyrite.
Siliceous ore (Keiko)	Stockwork, containing pyrite, chalcopyrite, and quartz.
Gypsum ore (Sekkoko)	A facies occurring adjacent to siliceous ore or yellow ore. Stratiform. Made up of mostly gypsum and anhydrite.

Source: From Lambert and Sato 1974, p. 1217.

There is an abundant literature on the Kuroko ores of Japan. For an overview, a good paper is Lambert and Sato (1974); two books are Tatsumi (1970) and Ishihara (1974). The ores have a characteristic stratigraphy (Fig. 8-14; Table 8-13), which, although varying greatly in the degree of development of the various members, is found in all deposits. From a sedimentary viewpoint, the ferruginous cherts are perhaps the most interesting facies, but little has been published about their composition, making comparisons with other iron accumulations difficult. Mineralogy and chemistry (Table 8-13, 8-14) are more complex than in Cyprus-type ores; note the prominence of Ba and Pb.

Table 8-14. Average chemical composition of Kuroko ores, in percent.

	Black Ore	Yellow Ore	Siliceous Ore
SiO_2	1.78	0.76	48.5
Al_2O_3	0.37	0.42	0.81
MgO	0.06	0.07	0.08
CaO	0.02	0.11	0.12
BaO	37.1	0.28	0.29
S (total)	24.4	49.8	24.4
Fe	2.50	42.8	21.6
Cu	1.41	5.38	2.24
Pb	9.42	0.27	0.48
Zn	18.3	0.15	0.16
As	0.12	0.04	0.04
Au (ppm)	2.1	0.7	0.4
Ag (ppm)	646	27	18

Source: Oshima and others 1974, table 5.

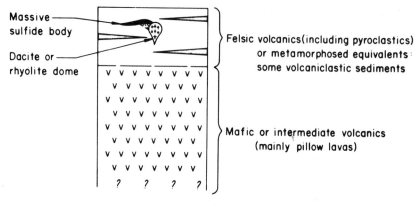

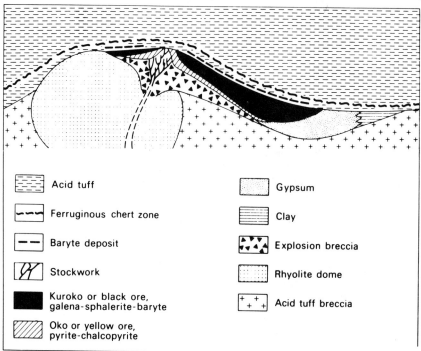

Fig. 8-14. Geologic framework of Kuroko deposits of Japan. Total thickness of the section in the top figure is 10–15,000 m (Sawkins 1976a, fig. 1). In the detailed cross-section, the orebody is typically 200–400 m across, as shown in the plan view (Evans 1980, figs. 13.7 and 13.8).

Fig. 8-14. (Continued)

In addition to the distinctive facies pattern, these deposits show a close association with felsic volcanism in the form of small dacitic or rhyodacitic domes (Fig. 8-14). Further, they are quite restricted in time-stratigraphic occurrence: almost all orebodies were emplaced at the end of the Nishikurosawa Stage of the Middle Miocene (Fig. 8-15). This was a time of waning volcanic activity and maximum submergence. Some individual deposits appear to have formed in small depressions on the sea floor, which could have restricted circulation, and favored preservation of the sulfides. Sato (1972) discussed various modes of deposition from the exiting fluid onto the sea floor.

Isotopic data suggest that, as in the mid-ocean ridge deposits, heated seawater was the hydrothermal fluid. Hydrogen and oxygen isotopes of water in fluid inclusions (Ohmoto and Rye 1974) are close to seawater values, and admit of only a small contribution from juvenile or meteoric sources, perhaps 10 percent. In fact, all of the variation seen can as easily be explained by reaction of seawater with the volcanics. Sulfur isotopes again indicate a seawater source, particularly for barite and anhydrite, which have a narrow range in $\delta^{34}S$, $+21$ to $+24$ permil (Fig. 8-16), close to seawater values. Oxygen in barite suggests that it formed from seawater at a temperature of 200-250°C (Sakai and Matsubaya 1971, p. 81); the isotopically heavier oxygen in the calcium sulfates could be a result of a different fractionation or of a lower temperature of crystallization. $\delta^{34}S$ of

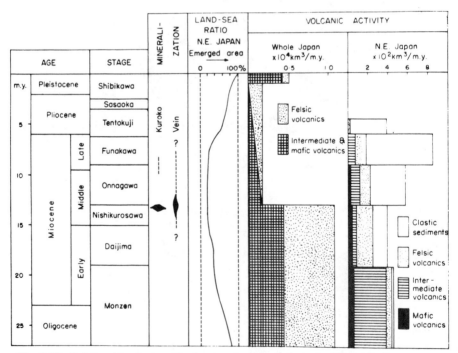

Fig. 8-15. Relationship of mineralization to volcanic activity in the "Green Tuff orogeny" of Japan (Lambert and Sato 1974, fig. 5).

chalcopyrite-pyrite pairs suggest temperatures of 250-300°C for the lower part of the deposits, 250°C for the barite-rich black ores (Fig. 8-17). Similarly, filling temperatures of fluid inclusions in quartz and sphalerite fall mainly in the range 200-250°C (Lambert and Sato 1974, p. 1225). Note also in Fig. 8-17 the positive S values, compared with mantle-derived S, again consistent with a seawater

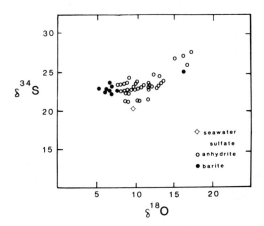

Fig. 8-16. Sulfur and oxygen isotopes of sulfates from Kuroko deposits suggest that seawater was the hydrothermal fluid and supplied the sulfur. (Data of Sakai and Matsubaya 1971, fig. 1.)

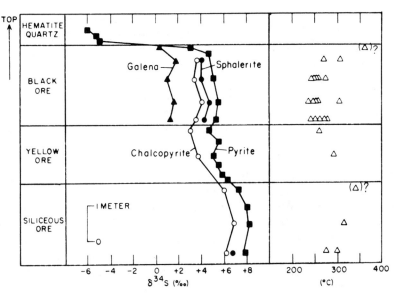

Fig. 8-17. Sulfur isotopes of sulfide minerals in Kuroko deposits suggest deposition at 250–300°C in a seawater-hydrothermal system, except for the ferruginous chert, which has biogenic sulfide (Rye and Ohmoto 1974, fig. 11).

source, and the abrupt shift to negative values in the overlying ferruginous chert, indicating bacterial sulfate reduction. Contrary to this model, however, is the contention of Urabe and Sato (1978) that, because the fluid inclusions contain water with a salinity higher than seawater, seawater was not the hydrothermal fluid, instead it was largely magmatic.

Kuroko-type deposits have been reported from other localities in the Pacific, for example from Fiji (Colley and Rice 1975). They are also known from older orogenic belts, two examples being Buchans in Newfoundland and some of the deposits of the Bathurst area in New Brunswick. At Buchans (Thurlow and others 1975, Swanson and others 1981), ores of similar composition to those of Japan, but lacking calcium sulfates, occur in a felsic pyroclastic sequence. Altered domes of white rhyolite are common, but lack the intimate association with ore seen in the Kuroko district (Thurlow and Swanson 1981, p. 137). Stratigraphic zoning is present, but not so well developed: siliceous stockwork ore is overlain by massive sulfides that increase upwards in Ba content, but the division into yellow and black ore cannot be made. Instead, the two ore types are closely interbedded or even mixed by slumping into "streaky ore." Also present are reworked orebodies made up of mechanically transported fragmental ore, probably carried downslope by debris flows, that maintain ore grades for distances of up to 2 km from their source in *in situ* massive sulfides (Thurlow and Swanson 1981). Iron-formation is not well developed at Buchans itself, but it is a prominent facies in related deposits of the same district (Dean 1978, p. 161). Stable isotopes (Kowalik and others 1981) indicate a seawater source for the S in a system with a large water/rock ratio depositing sphalerite and galena at 150 to 325°C. $\delta^{34}S$ for sulfides ranges from $+3$ to $+9$ permil, for barite from $+23$ to $+25$, almost identical to Kuroko, but somewhat heavier when corrected for Silurian seawater SO_4^{2-} of $+24$ permil, 4 permil heavier than in the Miocene, suggesting that f_{O_2} was higher than at Kuroko (Kowalik and others 1981, p. 250).

The situation at Bathurst is more complex, and has been obscured by metamorphism and tectonic disturbance. Over twenty orebodies have been found in the district, and appear to be of two kinds (Harley 1979). The first, which is economically more important, occurs as large strata-bound lenses in rhyolitic ash-flow tuffs with laterally equivalent banded iron-formation. Examples are Brunswick no.12, Heath Steele, and Orvan Brook. For size, the Orvan Brook orebody is 2.5 km long and up to 5 m thick (Harley 1979, p. 789), a considerably wider lateral extent than the Kuroko ores. The second type is smaller and more like a Kuroko-type deposit, being associated with brecciated felsic domes. An example is Stratmat West-North Boundary. The volcanics were formed in a mature island arc sequence and overlie continental clastics (Whitehead and Goodfellow 1978), thus conforming to the Kuroko pattern. Harley (1979) has suggested that the deposits formed during the development of a resurgent caldera, with Type 1 orebodies preceding Type 2. Unlike Buchans or the Kuroko deposits, barite is not reported, nor is there any mention of mineral zoning. Thus, although similar in many ways, the Kuroko ores are not exact analogues for these deposits.

In particular, it is puzzling why Type 1 deposits are not also found in Japan, considering the similarity in geologic setting.

Average $\delta^{34}S$ of S in the deposit is $+13.5$ with a range of $+10$ to $+16$ (Dechow 1960, Lusk 1972), with pyrite, chalcopyrite, sphalerite, and galena all having the same ranges. These values are about 9 permil heavier than S in the massive sulfide portion of the Kuroko deposits (Fig. 8-17), but, because Ordovician seawater was 7 permil heavier (Table 8-5) than in the Miocene, the two deposits are actually quite similar in $\delta^{34}S$. Bathurst S is 4.5 permil heavier than the roughly coeval ophiolite deposits of Notre Dame Bay, Newfoundland (Table 8-5), suggesting that physical conditions were somewhat different, perhaps temperature or f_{O_2}.

Before the large massive sulfides were discovered, the associated BIF was mined. It consists of three facies: magnetite, hematite, and chlorite (Boyle and Davies 1964). The magnetite facies, the largest, comprises alternating layers of chert and magnetite with hematite, chlorite, and siderite. The magnetite facies, in places, grades upwards and outwards into the hematite facies, similar in composition to the magnetite facies but with 40 percent specular hematite, 30 percent magnetite. Where undisturbed, this facies exhibits good banding, with laminae from 0.2 to 1.5 mm thick. Commonly the bands show complex slump-like folds suggesting penecontemporaneous deformation. The chlorite facies is gradational between the magnetite facies and sericite schists of the country rock. It contains fine bands of iron-rich chlorite (daphnite and prochlorite) and sericite, siderite, quartz, and calcite. The similarity of this iron-formation to Algoma-type BIF is striking, and shows that the volcanic-related iron-formations are not time-restricted in the way that the Lake Superior type BIF is. The Bathurst BIF deserves more study for comparison and contrast with other types.

In summary, deposits of convergent boundaries share many features with the mid-ocean ridge deposits. The process of ore formation—hydrothermal circulation of seawater through hot volcanic rocks with precipitation of the leached metals syngenetically on the sea floor—is the same. Temperatures are similar, as are the S isotopes. The major difference seems to be the presence of more metals in the island-arc deposits, but even this distinction breaks down for primitive island arcs.

Appendix

Free energy values for substances appearing in Eh–pH diagrams, in kilojoules.

Formula		Free Energy	Source
Dissolved Species			
Na^+		−261.90	1
K^+		−282.49	1
Mg^{++}		−454.80	1
Ca^{++}		−553.54	1
Fe^{++}		−78.87	1
Fe^{3+}		−4.60	1
Mn^{++}		−228.00	1
Cu^+		50.00	1
Cu^{++}		65.52	1
Ag^+		77.08	1
Ni^{++}		−45.60	1
Al^{3+}		−489.40	2
$(Al(OH)_2)^+$		−914.20	2
OH^-		−157.33	1
HCO_3^-		−586.85	1
HS^-		12.10	1
Cl^-		−131.27	1
$(H_3SiO_4)^-$		−1250.99	1
$(Al(OH)_4)^-$		−1305.00	2
CO_3^{--}		−527.90	1
SO_4^{--}		−744.63	1
S^{--}		85.80	1
CO_2 (g)		−394.38	1
H_2S (g)		−33.54	1
H_2S (aq)		−27.87	10
H_2O		−237.14	1
H_4SiO_4		−1307.50	2
H_2CO_3		−623.17	1
Aluminum minerals			
	$\gamma\text{-}Al_2O_3$	−1562.70	1
boehmite	$\gamma\text{-}AlOOH$	−918.4	2
gibbsite	$\gamma\text{-}Al(OH)_3$	−1154.9	2
corundum	$\alpha\text{-}Al_2O_3$	−1582.2	2

Formula		Free Energy	Source
diaspore	α-AlOOH	-922.0	2
bayerite	α-Al(OH)$_3$	-1153.0	2

Carbonates

calcite	CaCO$_3$	-1128.84	1
aragonite	CaCO$_3$	-1127.79	1
dolomite	CaMg(CO$_3$)$_2$	-2161.67	1
siderite	FeCO$_3$	-667.09	3
rhodochrosite	MnCO$_3$	-816.05	1
cerussite	PbCO$_3$	-625.34	1
smithsonite	ZnCO$_3$	-731.48	1

Oxides, hydroxides

hematite	Fe$_2$O$_3$	-742.68	1
magnetite	Fe$_3$O$_4$	-1012.57	1
goethite	FeOOH	-488.55	1
ferric hydroxide	Fe(OH)$_3$	-693.73	4
"green rust"	Fe$_3$(OH)$_8$	-1910.34	4
fresh	Fe(OH)$_3$	-688.33	3
Fe(OH)$_3$ aged 4 days		-695.15	3
bunsenite	NiO	-211.58	1
	Ni(OH)$_2$	-459.10	4
manganosite	MnO	-362.90	1
pyrolusite	βMnO$_2$	-465.14	1
bixbyite	Mn$_2$O$_3$	-881.07	1
pyrochroite	Mn(OH)$_2$	-616.91	8
hausmannite	Mn$_3$O$_4$	-1281.93	8
manganite	MnOOH	-558.03	8
birnessite	δ-MnO$_2$	-453.15	8
nsutite	γ-MnO$_2$	-456.62	8
bixbyite	α-Mn$_2$O$_3$	-878.03	9
tenorite	CuO	-129.56	1
cuprite	Cu$_2$O	-146.30	1
trevorite	NiFe$_2$O$_4$	-972.94	1

Sulfides

pyrite	FeS$_2$	-162.34	5
marcasite	FeS$_2$	-158.58	5
acanthite	Ag$_2$S	-40.08	1
covellite	CuS	-49.08	1
chalcocite	Cu$_2$S	-86.87	1
chalcopyrite	CuFeS$_2$	-187.87	6
bornite	Cu$_5$FeS$_4$	-362.78	6
albandite	MnS	-218.16	1
galena	PbS	-96.08	1
sphalerite	ZnS	-202.50	1

Formula		Free Energy	Source
Halides			
chlorargyrite	AgCl	-109.82	1
Silicates			
kaolinite	$Al_2Si_2O_5(OH)_4$	-3799.364	1
microcline	$KAlSi_3O_8$	-3742.33	1
muscovite	$KAl_3Si_3O_{10}(OH)_2$	-5600.671	1
talc	$Mg_3Si_4O_{10}(OH)_2$	-5536.048	1
chrysotile	$Mg_3Si_2O_5(OH)_4$	-4034.024	1
quartz	SiO_2	-856.29	1

Sources:
1. Robie and others 1978.
2. Hemingway and Robie 1978.
3. Langmuir 1979.
4. Calculated using method of Tardy and Garrels 1976.
5. Gronvold and Westrum 1976.
6. Helgeson and others 1978.
8. Recalculated from Bricker 1965 using free energy values for dissolved species from (1).
9. Crerar and others 1981.
10. Rose 1976.

References

Adams, S.S., H.S. Curtis, P.L. Hafen, and H. Salek-Nejad. 1978. Interpretation of postdepositional processes related to the formation and destruction of the Jackpile-Postgate uranium deposit, northwest New Mexico: Econ. Geol., v. 73, p. 1635-1654.

Adler, H. 1974. Concepts of uranium-ore formation in reducing environments in sandstones and other sediments. *In:* Formation of Uranium Deposits. Intl. Atomic Energy Agency, Vienna, p. 141-168.

Ahmad, N., R.L. Jones, and A.H. Beavers. 1966. Genesis, mineralogy and related properties of West Indian soils. Part 1, Bauxitic soils of Jamaica: Soil Sci. Soc. Amer. Proc. v. 30, p. 719-722.

Alling, H.L. 1947. Diagenesis of the Clinton hematite ores of New York: Geol. Soc. Amer. Bull., v. 58, p. 991-1018.

Amstutz, G.C. 1959. Syngenese und Epigenese in Petrographie und Lagerstättenkunde: Schweiz. Min. Petrology Mitt., v. 39, p. 1-84.

Amstutz, G.C. 1977. Time- and strata-bound features of the Michigan copper deposits. *In:* D.D. Klemm and H-J. Schneider, Eds., Time- and Strata-Bound Ore Deposits. Springer-Verlag, New York, Heidelberg, Berlin,, p. 123-140.

Anderson, C.A. 1955. Oxidation of copper sulfides and secondary sulfide enrichment: Econ. Geol., 50th Anniv. Volume, p. 324-340.

Anderson, G.M. 1975. Precipitation of Mississippi Valley type ores: Econ. Geol., v. 70, p. 937-942.

Anderson, R.Y., and D.W. Kirkland. 1966. Intrabasin varve correlation: Geol. Soc. Amer. Bull., v. 77, p. 241-256.

Andersson, A., B. Dahlman, and D.G. Gee. in press. Kerogen and uranium resources in the Cambrian alum shales of the Billingen-Falbygden and Narke areas, Sweden: Geologiska Föreningens Förhandlingar.

Anderton, R., P.H. Bridges, M.R. Leeder, and B.W. Sellwood. 1979. A Dynamic Stratigraphy of the British Isles. George Allen and Unwin, London, 301 pp.

Anger, G., H. Nielsen, H. Puchelt, and W. Ricke. 1966. Sulfur isotopes in the Rammelsberg ore deposit (Germany): Econ. Geol., v. 61, p. 511-536.

Anhaeusser, C.R., and A. Button. 1972. A petrographic and mineragraphic study of the copper-bearing formations in the Witvlei area, Southwest Africa: University Witwatersrand, Econ. Geol. Research Unit, Information Circ. 66, 32 pp.

Annels, A.E. 1974. Some aspects of the stratiform ore deposits of the Zambian copperbelt and their genetic significance. *In:* P. Bartholomé, Ed., Gisements Stratiformes et Provinces Cuprifères. Soc. géol. Belgique, Liège, p. 203-213.

Annels, A.E. 1979. Mufulira greywackes and their associated sulphides: Trans. Inst. Mining Metal., v. 88, p. B15-B22. Discussion, p. B190-B198.

Armands, G. 1972. Geochemical studies of uranium, molybdenum and vanadium in the Swedish Alum Shale: Stockholm Contr. Geol., v. 22, p. 1-148.

Armstrong, F.C., Ed. 1981. Genesis of Uranium- and Gold-Bearing Precambrian Quartz-Pebble Conglomerates. U.S. Geol. Survey, Prof. Paper 1161, various paging.

Ashraf, M., N.A. Chohan, and F.A. Faruqi. 1972. Bauxite and clay deposits in the Kattha area, Salt Range, Punjab, West Pakistan: Econ. Geol., v. 67, p. 103-110.

Augustithis, S.S., Ed. 1981. International Symposium on Metallogeny of Mafic and Ultramafic Complexes. v. I: Chromites and Laterites, UNESCO, Athens, 193 pp.

Awramik, S.M. 1976. Gunflint stromatolites: microfossil distribution in relation to stromatolite morphology. In: M.R. Walter, Ed., Stromatolites. Elsevier, Amsterdam. p. 311-320.

Awramik, S.M., and E.S. Barghoorn. 1977. The Gunflint microbiota: Precambrian Res., v. 5, p. 721-742.

Ayres, D.E. 1972. Genesis of iron-bearing minerals in banded iron formation mesobands in the Dales Gorge Member, Hamersley Group, Western Australia: Econ. Geol., v. 67, p. 1241-1233.

Ayres, D.E., and P.J. Eadington. 1975. Uranium mineralization in the South Alligator River valley: Min. Depos., v. 10, p. 27-41.

Bachinski, D.J. 1977. Sulfur isotopic composition of ophiolitic cupriferous iron sulfide deposits, Notre Dame Bay, Newfoundland: Econ. Geol., v. 72, p. 243-257.

Badham, J.P.N. 1981. Shale-hosted Pb-Zn deposits: products of exhalation of formation waters?: Trans. Inst. Mining Metal., v. 90, p. B70-B76.

Badham, J.P.N., and C.W. Stanworth. 1977. Evaporites from the Lower Proterozoic of the East Arm, Great Slave Lake: Nature, v. 268, p. 516-518.

Bailey, S.W., and S.A. Tyler. 1960. Clay minerals associated with the Lake Superior iron ores: Econ. Geol., v. 55, p. 150-173.

Balzer, W. 1982. On the distribution of iron and manganese at the sediment/water interface: thermodynamic versus kinetic control: Geochim. Cosmochim. Acta, v. 46, p. 1153-1161.

Bárdossy, G. 1982. Karst Bauxites. Elsevier, Amsterdam, 441 pp.

Bárdossy, G., and J.L. White. 1979. Carbonate inhibits the crystallization of aluminum hydroxide in bauxite: Science, v. 203, p. 355-356.

Bárdossy, G., A. Csanády, and A. Csordás. 1978. Scanning electron microscope study of bauxites of different ages and origins: Clays Clay Min., v. 26, p. 245-262.

Barghoorn, E.S. 1971. The oldest fossils: Scientific Amer., v. 244, p. 30-41.

Barghoorn, E.S., and S.A. Tyler. 1965. Microorganisms from the Gunflint Chert: Science, v. 147, p. 563-577.

Barghoorn, E.S., A.H. Knoll, H. Dembicki, and W.G. Meinschein. 1977. Variation in stable carbon isotopes in organic matter from the Gunflint Iron Formation: Geochim. Cosmochim. Acta, v. 41, p. 425-430.

Barnhisel, R.I., and C.I. Rich. 1965. Gibbsite, bayerite, and nordstrandite as affected by anions, pH and mineral surfaces: Soil Sci. Soc. Amer. Proc., v. 29, p. 531-534.

Bartholomé, P. 1974. On the diagenetic formation of ores in sedimentary beds, with special reference to Kamoto, Shaba, Zaire. *In:* P. Bartholomé, Ed., Gisements Stratiformes et Provinces Cuprifères. Soc. géol. Belgique, Liège, p. 203-213.

Bartholomé, P., P. Evrard, F. Katakesha, J. Lopez-Ruiz, and M. Ngongo. 1973. Diagenetic ore-forming processes at Kamoto, Katanga, Republic of the Congo. *In:* G.C. Amstutz and A.J. Bernard, Eds., Ores in Sediments. Springer-Verlag, New York, Heidelberg, Berlin, p. 21-42.

Bates, T.F. 1960. Halloysite and gibbsite formation in Hawaii: Clays Clay Min., v. 9, p. 315-328.

Bayley, R.W., and H.L. James. 1973. Precambrian iron-formations of the United States: Econ. Geol., v. 68, p. 934-959.

Beales, F.W. 1975. Precipitation mechanisms for Mississippi Valley-type ore deposits: Econ. Geol., v. 70, p. 943-948.

Beaudoin, B., A. Lesavre, and H. Pélissonnier. 1976. Action des eaux superficielles dans le gisement de manganèse d'Imini (Maroc): Bull. Soc. géol. France, v. 18, p. 95-100.

Becker, R.H., and R.N. Clayton. 1972. Carbon isotopic evidence for the origin of a banded iron-formation in Western Australia: Geochim. Cosmochim. Acta, v. 36, p. 577-595.

Becker, R.H., and R.N. Clayton. 1976. Oxygen isotope study of a banded iron-formation, Hamersley Range, Australia: Geochim. Cosmochim. Acta, v. 40, p. 1153-1165.

Bell, R.T., and G.D. Jackson. 1974. Aphebian halite and sulfate indications in the Belcher Group NWT: Canadian Jour. Earth Sci., v. 11, p. 722-728.

Berge, J.W., K. Johansson, and J. Jack. 1977. Geology and origin of the hematite ores of the Nimba Range, Liberia: Econ. Geol., v. 72, p. 582-607.

Bergström, J. 1980. Middle and Upper Cambrian biostratigraphy and sedimentation in south central Jamtland, Sweden: Geologiska Föreningens Förhandlingar, v. 102, p. 373-376.

Bernal, J.O., D.R. Dasgupta, and A.L. MacKay. 1959. The oxides and hydroxides of iron and their structural interrelationships: Clay Min. Bull., v. 4, p. 15-30.

Bernard, A.J. 1973. Metallogenic processes of intra-karstic sedimentation. *In:* G.C. Amstutz and A.J. Bernard, Eds., Ores in Sediments. Springer-Verlag, New York, Heidelberg, Berlin, p. 43-58.

Berner, R.A. 1970. Sedimentary pyrite formation: Amer. Jour. Sci., v. 268, p. 1-23.

Berner, R.A. 1971. Principles of Chemical Sedimentology. McGraw-Hill, New York, 240 pp.

Berner, R.A. 1978. Sulfate reduction and the rate of deposition of marine sediments: Earth Plan. Sci. Let., v. 37, p. 492-498.

Berner, R.A. 1981. A new geochemical classification of sedimentary environments: Jour. Sed. Petrology, v. 51, p. 359-365.

Berner, R.A., T. Baldwin, and G. Holdren. 1979. Authigenic iron sulfides as paleosalinity indicators: Jour. Sed. Petrology, v. 49, p. 1345-1350.

Besset, F., and J. Coudray. 1978. Le comportement du nickel dans les processus d'altération des péridotites de Nouvelle Calédonie: Bull. Bureau Recherches Géologiques Minières, ser. 2, sect. 2, p. 207-223.

Beukes, N.J. 1973. Precambrian iron-formations of southern Africa: Econ. Geol., v. 68, p. 960-1004.

Bhattacharyya, D.P., and P.K. Kakimoto. 1982. Origin of ferriferous ooids: an SEM study of ironstone ooids and bauxite pisoids: Jour. Sed. Petrology, v. 52, p. 849-857.

Bignell, R.D., D.S. Cronan, and J.S. Tooms. 1976. Metal dispersion in the Red Sea as an aid to marine geochemical exploration: Trans. Inst. Mining Metal., v. 85, p. B273-B278.

Billings, G.K., S.E. Kesler, and S.A. Jackson. 1969. Relation of zinc-rich formation waters, northern Alberta, to the Pine Point ore deposit: Econ. Geol., v. 64, p. 385-391.

Binda, P.L., and J.R. Mulgrew. 1974. Stratigraphy of copper occurrences in the Zambian copperbelt. *In:* P. Bartholomé, Ed., Gisements Stratiformes et Provinces Cuprifères. Soc. géol. Belgique, Liège, p. 215-233.

Bischoff, J.L. 1969. The Red Sea geothermal deposits: their mineralogy, chemistry and genesis. *In:* E.T. Degens and D. Ross, Eds., Hot Brines and Recent Heavy Metal Deposits in the Red Sea. Springer-Verlag, New York, Heidelberg, Berlin, p. 348-401.

Bischoff, J.L., and F.W. Dickson. 1975. Seawater-basalt interaction at 200°C and 500 bars: implications for origin of sea-floor heavy metal deposits and regulation of seawater chemistry: Earth Plan. Sci. Let., v. 25, p. 385-397.

Bischoff, J.L., and D.Z. Piper, Eds. 1979. Marine Geology and Oceanography of the Pacific Manganese Nodule Province. Plenum Press, New York, 842 pp.

Bish, D.L. 1978. Anion-exchange in takovite: applications to other hydroxide minerals: Bull. Bureau Recherches Géologique Minières, ser. 2, sect. 2, p. 293-301.

Bish, D.L., and G.W. Brindley. 1977. A reinvestigation of takovite, a nickel aluminum hydroxy-carbonate of the pyroaurite group: Amer. Min., v. 62, p. 458-464.

Bjørlykke, K., and J.O. Englund. 1979. Geochemical response to Upper Precambrian rift basin sedimentation and Lower Paleozoic epicontinental sedimentation in south Norway: Chem. Geol., v. 27, p. 271-295.

Boast, A.M., M.L. Coleman, and C. Halls. 1981. Textural and stable isotope evidence for the genesis of the Tynagh base metal deposit, Ireland: Econ. Geol., v. 76, p. 27-55.

Bonatti, E. 1975. Metallogenesis at oceanic spreading centers: Ann. Rev. Earth Plan. Sci., v. 3, p. 401-431.

Bonatti, E., and O. Joensuu. 1966. Deep-sea iron deposit from the South Pacific: Science, v. 154, p. 643-645.

Bonatti, E., D.E. Fisher, O. Joensuu, H.S. Rydell, M. Beyth. 1972a. Iron-manganese-barium deposit from the Northern Afar Rift (Ethiopia): Econ. Geol., v. 67, p. 717-730.

Bonatti, E., J. Honnorez, O. Joensuu, and H. Rydell. 1972b. Submarine iron deposits from the Mediterranean Sea. *In:* D.J. Stanley, Ed., The Mediterranean Sea. Hutchinson and Ross, Stroudsburg, Pennsylvania, p. 701-710.

Bonatti, E., B-M. Guerstein-Honnorez, J. Honnorez, and C. Stern. 1976a. Copper-iron sulfide mineralizations from the equatorial Mid-Atlantic ridge: Econ. Geol., v. 71, p. 1515-1525.

Bonatti, E., M. Zerbi, R. Kay, and H. Rydell. 1976b. Metalliferous deposits from the Apennine ophiolites: Mesozoic equivalent of modern deposits from oceanic spreading centers: Geol. Soc. Amer. Bull., v. 87, p. 83-94.

Bonhomme, M.G., F. Gauthier-Lafaye, and F. Weber. 1982. An example of Lower Proterozoic sediments: the Francevillian in Gabon: Precambrian Research, v. 18, p. 87-102.

Boni, M., and G.C. Amstutz. 1982. The Permo-Triassic paleokarst ores of southwest Sardinia (Inglesiente-Sulcis). An attempt at a reconstruction of paleokarst conditions. *In:* G.C. Amstutz, A. El Goresy, G. Frenzel, C. Kluth, G. Moh, A. Wauschkuhn, and R.A. Zimmermann, Eds., Ore Genesis: The State of the Art. Springer-Verlag, New York, Heidelberg, Berlin, p. 73-82.

Borchert, H. 1952. Die Bildungsbedingungen mariner Eisenerzlagerstätten: Chemie der Erde, v. 16, p. 49-74.

Borchert, H. 1960. Genesis of marine sedimentary iron ores: Trans. Inst. Mining Metal., v. 69, p. B261-B279. Discussion, p. B530-B535.

Borchert, H. 1980. On the genesis of manganese ore deposits. *In:* I.M. Varentsov and Gy. Grasselly, Eds., Geology and Geochemistry of Manganese. E. Schweizerbart'sche Verlagsbuchhandlung, Stuttgart, v. 2, p. 45-60.

Boström, K. 1973. The origin and fate of ferromanganoan active ridge sediments: Stockholm Contrib. Geol., v. 27, p. 149-243.

Boström, K., and M.N.A. Peterson. 1966. Precipitates from hydrothermal exhalations on the East Pacific Rise: Econ. Geol., v. 61, p. 1258-1265.

Bouladon, J., and G. Jouravsky. 1952. Manganèse. *In:* Géologie des Gîtes Minéraux Marocains. 19th Intl. Geol. Cong., Algiers, p. 45-80.

Bouladon, J., and G. Jouravsky. 1956. Les gîtes de manganèse du Maroc. *In:* J.G. Reyna, Ed., Symposium on Manganese. 20th Intl. Geol. Cong., Mexico, p. 217-248.

Bowen, R., and A. Gunatilaka. 1977. Copper: Its Geology and Economics, John Wiley & Sons, New York-Toronto, 366 pp.

Boyle, R.W., and J.L. Davies. 1964. Geology of the Austin Brook and Brunswick No. 6 sulphide deposits, Gloucester County, New Brunswick: Geol. Survey Canada, Paper 63-24.

Boyle, E.A., J.M. Edmond, and E.R. Sholkovitz. 1977. The mechanism of iron removal in estuaries: Geochim. Cosmochim. Acta, v. 41, p. 1313-1324.

Braitsch, O. 1971. Salt Deposits: Their Origin and Composition. Springer-Verlag, New York, Heidelberg, Berlin, 297 pp.

Brewer, P.G., and D.W. Spencer. 1969. A note on the chemical composition of the Red Sea brines. *In:* E.T. Degens and D. Ross, Eds., Hot Brines and Recent Heavy Metal Deposits in the Red Sea. Springer-Verlag, New York, Heidelberg, Berlin, p. 174-179.

Bricker, O.P. 1965. Some stability relations in the system $Mn-O_2-H_2O$ at 250°C and one atmosphere total pressure: Amer. Min., v. 50, p. 1296-1354.

Bridge, J. 1952. Discussion: Correlation of aluminum hydroxide and age of bauxite deposits: Problems of Clay and Laterite Genesis. Amer. Inst. Mining Metal. Engineers, p. 212-214.

Brigo, L., L. Kostelka, P. Omenetto, H-J. Schneider, E. Schroll, O. Schulz, and I. Štrucl. 1977. Comparative reflections on four Alpine Pb-Zn deposits. *In:* D.D. Klemm and H-J. Schneider, Eds., Time- and Strata-Bound Ore Deposits. Springer-Verlag, New York, Heidelberg, Berlin,, p. 273-293.

Brindley, G.W. 1978. The structure and chemistry of hydrous nickel-containing silicate and aluminate minerals: Bull. Bureau Recherches Géologique Minières, ser. 2, sec. 2, p. 233-245.

Brindley, G.W. 1982. Chemical compositions of berthierines—a review: Clays Clay Min., v. 30, p. 153-155.

Brindley, G.W., and P.T. Hang. 1973. The nature of garnierites 1. Structures, chemical composition and color characteristics: Clays Clay Min., v. 21, p. 27-40.

Brongersma-Sanders, M. 1971. Origin of major cyclicity of evaporites and bituminous rocks: an actualistic model: Marine Geol., v. 11, p. 123-144.

Brookfield, M. 1973. The paleoenvironment of the Abbotsbury Ironstone (Upper Jurassic) of Dorset: Paleontology, v. 16, p. 261-274.

Brown, A.C. 1971. Zoning in the White Pine copper deposit, Ontonagon County, Michigan: Econ. Geol., v. 66, p. 543-573.

Brown, A.C. 1974. The copper province of northern Michigan, USA. *In:* P. Bartholomé, Ed., Gisements Stratiformes et Provinces Cuprifères. Soc. géol. Belgique, Liège, p. 317-330.

Brown, A.C. 1981. Stratiform copper deposits and pene-exhalative environments: Geol. Soc. Amer., National Meeting Abstracts, p. 418.

Brown, J.S. 1970. Mississippi Valley type lead-zinc ores: Min. Depos., v. 5, p. 103-119.

Bubenicek, L. 1964. Étude sédimentologique de minerai de fer oolithique de Lorraine. *In:* G.C. Amstutz, Ed., Sedimentology and Ore Genesis. Elsevier, Amsterdam, p. 113-122.

Bubenicek, L. 1968. Géologie des minerais de fer oolithiques: Min. Depos., v. 3, p. 89-108.

Burger, P.A. 1979. The Greenvale nickel laterite orebody. *In:* D.J.I. Evans, R.S. Shoemaker, and H. Veltman, Eds., International Laterite Symposium. Amer. Inst. Mining, Metal., and Petroleum Engineers, New York, p. 24-37.

Burnie, S.W., H.P. Schwartz, and J.H. Crocket. 1972. A sulfur isotopic study of the White Pine mine, Michigan: Econ. Geol., v. 67, p. 895-914.

Burns, D.J. 1961. Some chemical aspects of bauxite genesis in Jamaica: Econ. Geol., v. 56, p. 1297-1303.

Burns, R.G. 1970. Mineralogical Applications of Crystal Field Theory. Cambridge University Press, Cambridge, 224 pp.

Burns, R.G., and B. Brown. 1972. Nucleation and mineralogical controls on the composition of manganese nodules. *In:* D.R. Horn, Ed., Ferromanganese Deposits on the Ocean Floor. Lamont-Doherty Geol. Observatory, Palisades, New York, p. 51-61.

Burns, R.G., and V.M. Burns. 1977a. The mineralogy and crystal chemistry of deep-sea manganese nodules, a polymetallic resource of the twenty-first century: Phil. Trans. Royal Soc. London, v. 286A, p. 283-301.

Burns, R.G., and V.M. Burns. 1977b. Mineralogy. *In:* G.P. Glasby, Ed., Marine Manganese Deposits. Elsevier, Amsterdam, p. 185-248.

Button, A. 1976a. Iron-formation as an end member in carbonate sedimentary cycles in the Transvaal Supergroup, South Africa: Econ. Geol., v. 71, p. 193-201.

Button, A. 1976b. Transvaal and Hamersley Basins—review of basin development and mineral deposits: Min. Sci. Engineering, v. 8, p. 262-293.

Button, A. 1979. Algal concentration in the Sabi River, Rhodesia: depositional model for Witwatersrand carbon?: Econ. Geol., v. 74, p. 1876-1882.

Byers, C.W. 1977. Biofacies patterns in euxinic basins: a general model: Soc. Econ. Paleontol. Min., Spec. Pub. 25, p. 5-17.

Cahen, L. 1974. Geological background to the copper-bearing strata of southern Shaba, Zaire. *In:* P. Bartholomé, Ed., Gisements Stratiformes et Provinces Cuprifères. Soc. géol. Belgique, Liège, p. 57-77.

Caïa, J. 1976. Paleogeographical and sedimentological controls of copper, lead, and zinc mineralizations in the Lower Cretaceous sandstones of Africa: Econ. Geol., v. 71, p. 409-422.

Cailteux, J. 1974. Les sulfures du gisement cuprifère stratiforme de Musoshi, Shaba, Zaire. *In:* P. Bartholomé, Ed., Gisements Stratiformes et Provinces Cuprifères. Soc. géol. Belgique, Liège, p. 267-276.

Calhoun, T.A., and R.W. Hutchinson. 1981. Determination of flow direction and source of fragmental sulphides, Clementine Deposit, Buchans, Newfoundland. *In:* E.A. Swanson, D.F. Strong, and J.G. Thurlow, Eds., The Buchans Orebodies: Fifty Years of Geology and Mining. Geol. Assoc. Canada, Spec. Paper no. 22, p. 187-204.

Callendar, E., and C.J. Bowser. 1976. Freshwater ferromanganese deposits. *In:* K.H. Wolf, Ed., Handbook of Strata-Bound and Stratiform Ore Deposits. Elsevier, Amsterdam, v. 7, p. 341-394.

Calvert, S.E., and N.B. Price. 1970. Composition of manganese nodules and manganese carbonates from Loch Fyne, Scotland: Contr. Min. Petrology, v. 29, p. 215-233.

Calvert, S.E., and N.B. Price. 1977. Geochemical variation in ferromanganese nodules and associated sediments from the Pacific Ocean: Marine Chem., v. 5, p. 43-74.

Campbell, F.A., V.G. Ethier, H.R. Krouse, and R.A. Both. 1978. Isotopic composition of sulfur in the Sullivan orebody, British Columbia: Econ. Geol., v. 73, p. 246-268.

Campbell, N. 1967. Tectonics, reefs, and stratiform lead-zinc deposits of the Pine Point area, Canada. *In:* J.S. Brown, Ed., Genesis of Stratiform Lead-Zinc-Barite-Fluorite Deposits in Carbonate Rocks. Economic Geology Publishing Co., Lancaster, Pa., p. 59-70.

Cann, J.R., C.K. Winter, and R.G. Pritchard. 1977. A hydrothermal deposit from the floor of the Gulf of Aden: Min. Mag., v. 41, p. 193-199.

Cannon, W.F. 1976. Hard iron ore of the Marquette Range, Michigan: Econ. Geol., v. 71, p. 1012-1028.

Carpenter, A.B., M.L. Trout, and E.E. Pickett. 1974. Preliminary report on the origin and chemical evolution of lead- and zinc-rich oil field brines in central Mississippi: Econ. Geol., v. 69, p. 1191-1206.

Carroll, D. 1958. Role of clay minerals in the transportation of iron: Geochim. Cosmochim. Acta, v. 14, p. 1-27.

Cayeux, L. 1922. Les Minerais de Fer oolithiques de France. II. Minerais de Fer secondaires: Études des Gîtes Minéraux de France, Service de la Carte Géologique, Paris, 1052 pp.

Chapman, H.J., and E.T.C. Spooner. 1977. ^{87}Sr enrichment of ophiolitic sulphide deposits in Cyprus confirms ore formation by circulating seawater: Earth Plan. Sci. Let., v. 35, p. 71-78.

Chauvel, J-J. 1974. Les minerais de fer de l'Ordovicien inférieur du bassin de Bretagne-Anjou, France: Sedimentology, v. 21, p. 127-147.

Chauvel, J-J. and E. Dimroth. 1974. Facies types and depositional environments of the Sokoman iron formation, Labrador Trough, Quebec, Canada: Jour. Sed. Petrology, v. 44, p. 299-327.

Cheney, E.S. 1981. The hunt for giant uranium deposits: Amer. Scientist, v. 69, p. 37-48.

Cheney, E.S., and M.L. Jensen. 1966. Stable isotopic geology of the Gas Hills, Wyoming uranium district: Econ. Geol., v. 61, p. 44-71.

Cheney, E.S., and J.W. Trammell. 1973. Isotopic evidence for inorganic precipitation of uranium roll ore bodies: Amer. Assoc. Petroleum Geol. Bull., v. 57, p. 1297-1304.

Chowns, T.M., and F.K. McKinney. 1980. Depositional facies in Middle-Upper Ordovician and Silurian rocks of Alabama and Georgia. In: R.W. Frey, Ed., Excursions in Southeastern Geology. Geol. Soc. Amer. Annual Meeting, p. 323-348.

Chukrov, F.V., A.I. Gorshkov, I.V. Vitoskaya, V.A. Drits, and A.V. Sivtsov. 1982. On the nature of Co-Ni asbolane; a component of some supergene ores. In: G.C. Amstutz, A. El Goresy, G. Frenzel, C. Kluth, G. Moh, A. Wauschkuhn, and R.A. Zimmermann, Eds., Ore Genesis: The State of the Art. Springer-Verlag, New York, Heidelberg, Berlin, p. 230-239.

Clarke, O.M. 1966. The formation of bauxite on karst topography in Eufaula district, Alabama, and Jamaica, West Indies: Econ. Geol., v. 61, p. 903-916. Discussion, p. 1458-1459.

Claypool, G.E., W.T. Holser, I.R. Kaplan, H. Sakai, and I. Zak. 1980. The age curves of sulfur and oxygen isotopes in marine sulfate and their mutual interpretation: Chem. Geol., v. 28, p. 199-259.

Cloud, P. 1972. A working model for the primitive Earth: Amer. Jour. Sci., v. 272, p. 537-548.

Cluff, R.M. 1980. Paleoenvironments of the New Albany Shale Group (Devonian-Mississippian) of Illinois: Jour. Sed. Petrology, v. 50, p. 767-780.

Colley, H., and C.M. Rice. 1975. A Kuroko-type deposit in Fiji: Econ. Geol., v. 70, p. 1373-1386.

Collins, J.A., and L. Smith. 1973. Lithostratigraphic controls of some Ordovician sphalerite. *In:* G.C. Amstutz and A.J. Bernard, Eds., Ores in Sediments. Springer-Verlag, New York, Heidelberg, Berlin, p. 79-92.

Comer, J.B. 1974. Genesis of Jamaican bauxite: Econ. Geol., v. 69, p. 1251-1264. Discussion, v. 71, p. 821-823.

Conant, L.C., and V.E. Swanson. 1961. Chattanooga Shale and Related Rocks of Central Tennessee and Nearby Areas. U.S. Geol. Survey, Prof. Paper 357, 91 pp.

Constantinou, G., and G.J.S. Govett. 1972. Genesis of sulfide deposits, ochre, and umber of Cyprus: Trans. Inst. Mining Metal., v. B81, p. 34-46.

Constantinou, G., and G.J.S. Govett. 1973. Geology, geochemistry, and genesis of Cyprus sulfide deposits: Econ. Geol., v. 68, p. 843-858.

Corliss, J.B., M. Lyle, J. Dymond, and K. Crane. 1978. The chemistry of hydrothermal mounds near the Galapagos Rift: Earth Plan. Sci. Let., v. 40, p. 12-24.

Corliss, J.B., J. Dymond, L.I. Gordon, J.M. Edmond, R.P. vanHerzen, R.D. Ballard, K. Green, D. Williams, A. Bambridge, K. Crane, and T.H. vanAndel. 1979. Submarine thermal springs on the Galapagos Rift: Science, v. 203, p. 1073-1082.

Craig, H. 1966. Isotopic composition and origin of the Red Sea and Salton Sea brines: Science, v. 134, p. 1544.

Craig, H. 1969. Geochemistry and origin of the Red Sea brines, *In:* E.T. Degens and D. Ross, Eds., Hot Brines and Recent Heavy Metal Deposits in the Red Sea. Springer-Verlag, New York, Heidelberg, Berlin, p. 208-242.

Crerar, D.A., R.K. Cormick, and H.L. Barnes. 1980a. Geochemistry of manganese, an overview. *In:* I.M. Varentsov and Gy. Grasselly, Eds., Geology and Geochemistry of Manganese. E. Schweizerbart'sche Verlagsbuchhandlung, Stuttgart, v. 1, p. 293-334.

Crerar, D.A., A.G. Fischer, and C.L. Plaza. 1980b. *Metallogenium* and biogenic deposition of manganese from Precambrian to Recent time. *In:* I.M. Varentsov and Gy. Grasselly, Eds., Geology and Geochemistry of Manganese. E. Schweizerbart'sche Verlagsbuchhandlung, Stuttgart, v. 3, p. 285-304.

Cronan, D.S. 1974. Authigenic minerals in deep-sea sediments. *In:* E.D. Goldberg, Ed., The Sea. John Wiley & Sons, New York, p. 491-526.

Cronan, D.S. 1976. Implications of metal dispersion from hydrothermal systems for mineral exploration on mid-ocean ridges and in island arcs: Nature, v. 262, p. 567-569.

Cronan, D.S. 1980. Underwater Minerals. Academic Press, London, 362 pp.

Croxford, N.J.W., B.L. Gulson, and J.W. Smith. 1975. The McArthur Deposit: a review of the current situation: Min. Depos., v. 10, p. 302-304.

Cseh-Németh, J., J. Konda, Gy. Grasselly, and Z. Szabó. 1980. Sedimentary manganese deposits of Hungary. *In:* I.M. Varentsov and Gy. Grasselly, Eds., Geology and Geochemistry of Manganese. E. Schweizerbart'sche Verlagsbuchhandlung, Stuttgart, v. 2, p. 199-222.

Cuney, M. 1978. Geolgic environment, mineralogy, and fluid inclusions of the Bois Noirs-Limouzat Vein, Forez, France: Econ. Geol., v. 73, p. 1567-1610.

Curtis, C.D. 1967. Diagenetic iron minerals in some British Carboniferous sediments: Geochim. Cosmochim. Acta, v. 31, p. 2109-2123.

Curtis, C.D.. 1977. Sedimentary geochemistry, environments and processes dominated by involvement of an aqueous phase: Phil. Trans. Royal Soc. London, v. A286, p. 353-372.

Curtis, C.D., and D.A. Spears. 1968. The formation of sedimentary iron minerals: Econ. Geol., v. 63, p. 257-270.

Dahlkamp, F.J. 1978. Geologic appraisal of the Key Lake U-Ni deposits, northern Saskatchewan: Econ. Geol., v. 73, p. 1430-1449.

Davidson, C.F. 1966. Some genetic relationships between ore deposits and evaporites: Trans. Inst. Mining Metal., v. 75, p. B216-B225.

Davison, W. 1979. Soluble inorganic ferrous complexes in natural waters: Geochim. Cosmochim. Acta, v. 43, p. 1693-1696.

Dean, P.L. 1978. The Volcanic Stratigraphy and Metallogeny of Notre Dame Bay, Newfoundland. Memorial University of Newfoundland, Geology Report 7, 205 pp.

Dechow, E. 1960. Geology, sulfur isotopes and the origin of the Heath Steele ore deposits, Newcastle, N.B., Canada: Econ. Geol., v. 55, p. 539-556.

Dechow, E., and M.L. Jensen. 1965. Sulfur isotopes of some central African sulfide deposits: Econ. Geol., v. 60, p. 894-941.

Deer, W.A., R.A. Howie, and J. Zusman. 1962. The Rock-Forming Minerals. Longmans, London, v. 5, 365 pp.

Deffeyes, K.S. 1970. The axial valley: a steady-state feature of the terrain. In: H. Johnson and B.L. Smith, Eds., The Megatectonics of Continents and Oceans. Rutgers University Press, Camden, N.J., p. 194-222.

Degens, E.T., and G. Kulbicki. 1973. Hydrothermal origin of metals in some East African lakes: Min. Depos., v. 8, p. 388-404.

Degens, E.T., and D.A. Ross. 1976. Strata-bound metalliferous deposits found in or near active rifts. In: K.H. Wolfe, Ed., Handbook of Strata-Bound and Stratiform Ore Deposits. Elsevier, Amsterdam, v. 4, p. 165-202.

Degens, E.T., and P. Stoffers. 1977. Phase boundaries as an instrument for metal concentration in geological systems. In: D.D. Klemm and H-J. Schneider, Eds., Time and Strata-Bound Ore Deposits. Springer-Verlag, New York, Heidelberg, Berlin,, p. 25-45.

Degens, E.T., H. Okada, S. Honjo, and J.C. Hathaway. 1972. microcrystalline sphalerite in resin globules suspended in Lake Kivu, East Africa: Min. Depos., v. 7, p. 1-12.

Degens, E.T., R.P. vonHerzen, H.K. Wong, W.G. Deuser, and H.W. Jannasch. 1973. Lake Kivu: structure, chemistry, and biology of an East African rift lake: Geol. Rundschau, v. 61, p. 245-277.

Degens, E.T., F. Khoo, and W. Michaelis. 1977. Uranium anomaly in Black Sea sediments: Nature, v. 269, p. 566-569.

Dennen, W.H., and H.A. Norton. 1977. Geology and geochemistry of bauxite deposits in the lower Amàzon basin: Econ. Geol., v. 72, p. 82-89.

Derry, D.R., G.R. Clark, and N. Gillatt. 1965. The Northgate base metal deposit at Tynagh, County Galway, Ireland: Econ. Geol., v. 69, p. 17-25.

Deudon, M. 1957. Présence de maghemite dans le minerai de fer de Lorraine: Soc. française minéralogie et cristallographie Bull., v. 80, p. 239-241.

deVilliers, P.R. 1971. The geology and mineralogy of the Kalahari manganese-field north of Sishen, Cape Province: South Africa Geol. Survey, Dept. Mines, Memoir 59, 84 pp.

de Vries, J.J. 1974. Groundwater Flow Systems and Stream Nets in the Netherlands. Rodopi, N.V., Amsterdam, 226 pp.

di Colbertaldo, D. 1967. Génèse des gîtes minéraux à plomb-zinc dans les Alpes Centre-Orientales. *In:* J.S. Brown, Ed., Genesis of Stratiform Lead-Zinc-Barite-Fluorite Deposits in Carbonate Rocks. Economic Geology Publishing Co., Lancaster, Pa., p. 308-315.

Dimroth, E. 1976. Aspects of the sedimentary petrology of cherty iron-formation, *In:* K.H. Wolfe, Ed., Handbook of Strata-Bound and Stratiform Ore Deposits. Elsevier, Amsterdam, v. 7, p. 203-254.

Dimroth, E. 1977. Facies models 5. Diagenetic facies of iron formation: Geoscience Canada, v. 4, p. 83-88.

Dimroth, E., and J-J. Chauvel. 1973. Petrography of the Sokoman Iron Formation in part of the central Labrador Trough, Quebec, Canada: Geol. Soc. Amer. Bull., v. 84, p. 111-134.

Dimroth, E., and M.M. Kimberley. 1976. Precambrian atmospheric oxygen: evidence in the sedimentary distributions of carbon, sulfur, uranium, and iron: Canadian Jour. Earth Sci., v. 13, p. 1161-1185.

Dingess, P. 1976. Geology and mining operations at the Creta copper deposit of Eagle-Picher Industries, Inc.: Oklahoma Geol. Survey, Circular 77, p. 15-24.

Dodson, R.G., R.S. Needham, P.G. Wilkes, R.W. Page, P.G. Smart, and A.L. Watchmann. 1974. Uranium mineralization in the Rum Jungle-Alligator Rivers province, N.T., Australia. *In:* Formation of Uranium Deposits. Intl. Atomic Energy Agency, Vienna, p. 551-567.

Donnelly, T.W., and J.J.W. Rogers. 1978. The distribution of igneous rock suites throughout the Caribbean: Geol. Mijnbouw, v. 57, p. 151-162.

Donnelly, T.W., G. Thompson, and P.T. Robinson. 1979. Very low temperature alteration of the oceanic crust and the problem of fluxes of potassium and magnesium in deep drilling results in the Atlantic Ocean. *In:* M. Talwani, C.G. Harrison, and D.E. Hayes, Eds., Ocean Crust, Maurice Ewing Series No. 2, Amer. Geophysical Union, Washington, D.C., p. 369-381.

Donnelly, T.W., G. Thompson, and M.H. Salisbury. 1980. The chemistry of altered basalts at site 417, Deep-sea Drilling Project leg 51: Initial Reports, Deep-sea Drilling Project, v. 51-53, part 2, p. 1319-1330.

Dorr, J.V.N. 1964. Supergene iron ores of Minas Gerais, Brazil: Econ. Geol., v. 59, p. 1203-1241.

Drever, J.I. 1974. Geochemical model for the origin of Precambrian banded iron formations: Geol. Soc. Amer. Bull., v. 85, p. 1099-1106.

Duhovnik, J. 1967. Facts speaking for and against a syngenetic origin of the stratiform deposits of lead and zinc. *In:* J.S. Brown, Ed., Genesis of Stratiform Lead-Zinc-Barite-Fluorite Deposits in Carbonate Rocks. Economic Geology Publishing Co., Lancaster, Pa., p. 108-125.

Duke, N.A., and R.W. Hutchinson. 1974. Geological relationships between massive sulfide bodies and ophiolitic volcanic rocks near York Harbour, Newfoundland: Canadian Jour. Earth Sci., v. 11, p. 53-63.

Dunham, K.C. 1952. Petrography of the Liassic ironstones. *In:* T.H. Whitehead, W. Anderson, V. Wilson, and D.A. Wray, The Liassic Ironstones. Geol. Survey Great Britain, Memoir, p. 16-34.

Dunham, K.C. 1964. Neptunist concepts of ore genesis: Econ. Geol., v. 59, p. 1-21.

Dury, G.H. 1969. Rational descriptive classification of duricrusts: Earth Sci. Jour., v. 3, p. 77-86.

Eargle, D.H., and A.M.D. Weeks. 1973. Geologic relations among uranium deposits, South Texas, coastal plain region, U.S.A. *In:* G.C. Amstutz and A.J. Bernard, Eds., Ores in Sediments. Springer-Verlag, New York, Heidelberg, Berlin, p. 101-114.

Eargle, D.H., K.A. Dickinson, and B.O. Davis. 1975. South Texas uranium deposits: Amer. Assoc. Petroleum Geol. Bull., v. 59, p. 766-779.

Edmond, J.M., C. Measures, R.E. McDuff, L.H. Chan, R. Collier, B. Grant, L.I. Gordon, and J.B. Corliss. 1979. Ridge crest hydrothermal activity and the balances of the major and minor elements in the ocean: the Galapagos data: Earth Plan. Sci. Let., v. 46, p. 1-19.

Edzwald, J.K., and C.R. O'Melia. 1975. Clay distribution in recent estuarine sediments: Clays Clay Min., v. 23, p. 39-44.

Eggleton, R.A., and B.W. Chappell. 1978. The crystal structure of stilpnomelane. Part III: Chemistry and physical properties: Min. Mag., v. 42, p. 361-368.

Eichler, J. 1976. Origin of the Precambrian banded iron-formations. *In:* K.H. Wolfe, Ed., Handbook of Strata-Bound and Stratiform Ore Deposits. Elsevier, Amsterdam, v. 7, p. 157-202.

Eichmann, R., and M. Schidlowski. 1975. Isotopic fractionation between coexisting organic carbon-carbonate pairs in Precambrian sediments: Geochim. Cosmochim. Acta, v. 39, p. 585-595.

Elder, J.W. 1965. Physical processes in geothermal areas. *In:* W.H.K. Lee, Ed., Terrestrial Heat Flow. Amer. Geophysical Union, Monograph No. 8, p. 211-229.

Elderfield, H., I.G. Gass, A. Hamond, and L.M. Bear. 1972. The origin of ferromanganese sediments associated with the Troodos Massif of Cyprus: Sedimentology, v. 19, p. 1-19.

Elias, M., M.J. Donaldson, and N. Giorgetta. 1981. Geology, mineralogy, and chemistry of lateritic nickel-cobalt deposits near Kalgoorlie, Western Australia: Econ. Geol., v. 76, p. 1775-1783.

Emerson, S., R.E. Cranston, and P.S. Liss. 1979. Redox species in a reducing fjord: equilibrium and kinetic considerations: Deep-Sea Research, v. 26A, p. 859-878.

Ensign, C.O., W.S. White, J.C. Wright, J.L. Patrick, R.J. Leone, D.J. Hathaway, J.W. Trammell, J.J. Fritts, and T.L. Wright. 1968. Copper deposits in the Nonesuch Shale, White Pine, Michigan. *In:* J.D. Ridge, Ed., Ore Deposits of the United States. Amer. Inst. Mining Metal. Engineers, New York, v. 1, p. 460-488.

Ervin, G., and E.F. Osborn. 1951. The system Al_2O_3-H_2O: Jour. Geol., v. 59, p. 381-394.

Espenshade, G.H. 1954. Geology and Mineral Deposits of the James River-Roanoke River Manganese District, Virginia. U.S. Geol. Survey, Bull. 1008, 155 pp.

Esson, J., and L. Carlos. 1978. The occurrence, mineralogy, and chemistry of some garnierites from Brazil: Bull. Bureau Recherches Géologique Minières, ser. 2, sec. 2, p. 263-274.

Ethier, V.G., F.A. Campbell, R.A. Both, and H.R. Krouse. 1976. Geological setting of the Sullivan orebody and estimates of temperatures and pressure of metamorphism: Econ. Geol., v. 71, p. 1570-1588.

Eugster, H.P. 1969. Inorganic bedded cherts from the Magadi area, Kenya: Contr. Min. Petrology, v. 22, p. 1-31.

Eugster, H.P., and I-Ming Chou. 1973. The depositional environments of Precambrian iron-formations: Econ. Geol., v. 68, p. 1144-1168.

Eupene, G.S. 1980. Stratigraphic, structural, and temporal control of mineralization in the Alligator Rivers uranium province, Northern Territory, Australia, *In:* J.D. Ridge, Ed., Proceedings of the Fifth Quadrennial IAGOD Symposium. E. Schweizerbart's-che Verlagsbuchhandlung, Stuttgart, p. 347-376.

Evans, A.M. 1976. Genesis of Irish base-metal deposits. *In:* K.H. Wolfe, Ed., Handbook of Strata-Bound and Stratiform Ore Deposits. Elsevier, Amsterdam, v. 5, p. 231-256.

Evans, A.M. 1980. An Introduction to Ore Geology. Elsevier, New York, 231 pp.

Ewers, W.E., and R.C. Morris. 1981. Studies of the Dales Gorge Member of the Brockman Iron Formation, Western Australia: Econ. Geol., v. 76, p. 1929-1953.

Faust, G.T. 1966. The hydrous nickel-magnesium silicates— the garnierite group: Amer. Min., v. 51, p. 279-298.

Feather, C.E. 1981. Some aspects of Witwatersrand mineralization with special reference to uranium minerals. *In:* F.C. Armstrong, Ed., Genesis of Uranium- and Gold-Bearing Precambrian quartz-pebble conglomerates. U.S. Geol. Survey, Prof. Paper 1161, p. Q1-Q23.

Feather, C.E., and G.M. Koen. 1975. The mineralogy of the Witwatersrand reefs: Min. Sci. Engineering, v. 7, p. 189-224.

Feitnecht, von W. 1959. Über die Oxydation von festen Hydroxyverbindungen des Eisens in wäserigen Losungen: Zeits. für Elektrochemie, v. 63, p. 34-43.

Feitnecht, von W., and G. Keller. 1950. Über die dunkel grunen Hydroxyverbindungen des Eisens: Zeits. anorganische Chemie, v. 262, p. 61-68.

Ferguson, J., and I.B. Lambert. 1972. Volcanic exhalations and metal enrichments at Matupi Harbor, T.P.N.G: Econ. Geol., v. 67, p. 25-37.

Ferguson, J., I.B. Lambert, and H.E. Jones. 1974. Iron sulphide formation in an ex-halative-sedimentary environment, Talasea, New Britain, P.N.G: Min. Depos., v. 9, p. 33-47.

Finlow-Bates, T. 1980. The chemical and physical controls on the genesis of submarine exhalative orebodies and their implications for formulating exploration concepts. A review: Geol. Jahrbuch, Reihe D, v. 40, p. 131-168.

Fischer, R.P. 1970. Similarities, differences, and some genetic problems of the Wyoming and Colorado Plateau types of uranium deposits in sandstone: Econ. Geol., v. 65, p. 778-784.

Fleet, A.J., and A.H.F. Robertson. 1980. Ocean-ridge metalliferous and pelagic sediments of the Semail Nappe, Oman: Jour. Geol. Soc., v. 137, p. 403-422.

Fleischer, V.D., W.G. Garlick, R. Haldane. 1976. Geology of the Zambian copperbelt. *In:* K.H. Wolf, Ed., Handbook of Strata-Bound and Stratiform Ore Deposits. Elsevier, Amsterdam, v. 6, p. 223-352.

Floran, R.J., and J.J. Papike. 1975. Petrology of low-grade rocks of the Gunflint Iron Formation, Ontario-Minnesota: Geol. Soc. Amer. Bull., v. 86, p. 1169-1190.

Folk, R.L. 1962. Petrography and origin of the Silurian Rochester and McKenzie Shales, Morgan County, West Virginia: Jour. Sed. Petrology, v. 32, p. 539-578.

Folk, R.L., and J.S. Pittman. 1971. Length-slow chalcedony: a new testament for vanished evaporites: Jour. Sed. Petrology, v. 41, p. 1045-1058.

Francheteau, J., H.D. Needham, P. Choukraune, T. Juteau, M. Seguret, R.D. Ballard, P.J. Fox, W. Normark, A. Carranza, D. Cordoba, J. Guerro, C. Rangin, H. Bougault, P. Cambon, and R. Hékinian. 1979. Massive deep-sea sulfide ore deposits discovered by submersible on the East Pacific Rise: Project RITA, 21°N: Nature, v. 277, p. 523-528.

François, A. 1974. Stratigraphie, tectonique et minéralisations dans l'arc cuprifère du Shaba (Republique du Zaire). *In:* P. Bartholomé, Ed., Gisements Stratiformes et Provinces Cuprifères. Soc. géol. Belgique, Liège, p. 79-101.

Franklin, J.M., J.W. Lydon, and D.F. Sangster. 1981. Volcanic-associated massive sulfide deposits: Econ. Geol., 75th Anniv. Vol., p. 485-627.

French, V.M. 1968. Progressive contact metamorphism of the Biwabik Iron-formation, Mesabi Range, Minnesota: Minnesota Geol. Survey, Bull. 45, p. 3-103.

French, V.M. 1973. Mineral assemblages in diagenetic and low-grade metamorphic iron-formation: Econ. Geol., v. 68, p. 1063-1074.

Frenzel, G. 1980. The manganese ore minerals. *In:* I.M. Varentsov and Gy. Grasselly, Eds., Geology and Geochemistry of Manganese. E. Schweizerbart'sche Verlagsbuchhandlung, Stuttgart, v. 1, p. 25-158.

Frietsch, R.H., H. Papunen, and F.M. Vokes. 1979. The ore deposits of Finland, Norway, and Sweden: a review: Econ. Geol., v. 74, p. 975-1001.

Fritz, P., P.L. Binda, F.E. Follinsbee, and H.R. Krouse. 1971. Isotopic composition of diagenetic siderites from Cretaceous sediments, Western Canada: Jour. Sed. Petrology, v. 41, p. 282-288.

Froese, E. 1981. Applications of thermodynamics in the study of mineral deposits: Geol. Survey Canada, Paper 80-28, 38 pp.

Frost, B.R. 1978. Some aspects of the sedimentary and diagenetic environment of Proterozoic banded iron-formations—a discussion: Econ. Geol., v. 73, p. 1369-1373.

Frost, B.R. 1979. Metamorphism of iron-formation: parageneses in the system Fe-Si-C-O-H: Econ. Geol., v. 74, p. 775-785.

Fryer, B.J. 1977a. Trace element geochemistry of the Sokoman Iron Formation: Canadian Jour. Earth Sci., v. 14, p. 1598-1610.

Fryer, B.J. 1977b. Rare earth evidence in iron-formations for changing Precambrian oxidation states: Geochim. Cosmochim. Acta, v. 41, p. 361-367.

Furbish, W.J., and E.L. Schrader. 1977. Secondary biogenic textures in some iron-manganese nodules from the Blake Plateau, Atlantic Ocean: Marine Geol., v. 25, p. 343-354.

Gabelman, J.W. 1977. Migration of U and Th; exploration significance: Amer. Assoc. Petroleum Geol., Studies in Geology#3, 168 pp.

Galloway, W.E. 1977. Catahoula Formation of the Texas Coastal Plain: depositional systems, composition, structural development, ground-water flow history, and uranium distribution: University Texas at Austin, Bureau Econ. Geol., Report of Investigations no. 87, 59 pp.

Galloway, W.E. 1978. Uranium mineralization in a coastal-plain fluvial aquifer system: Catahoula Formation, Texas: Econ. Geol., v. 73, p. 1655-1676.

Galloway, W.E., and W.R. Kaiser. 1980. Catahoula Formation of the Texas coastal plain: origin, geochemical evolution, and characteristics of the uranium deposits: University Texas at Austin, Bureau Econ. Geol., Report of Investigations no. 100, 81 pp.

Gardner, L.R. 1980. Mobilization of Al and Ti during weathering—isovolumetric geochemical evidence: Chem. Geol., v. 30, p. 151-165.

Garlick, W.G. 1961. The syngenetic theory. In: F. Mendelsohn, Ed., The Geology of the Northern Rhodesian Copperbelt. McDonald, London, p. 146-165.

Garlick, W.G. 1964. Association of mineralization and algal reef structures on Northern Rhodesian copperbelt, Katanga, and Australia: Econ. Geol., v. 59, p. 416-427.

Garlick, W.G. 1981. Sabkhas, slumping, and compaction at Mufulira, Zambia: Econ. Geol., v. 76, p. 1817-1832.

Garlick, W.G., and V.D. Fleischer. 1972. Sedimentary environment of Zambian copper deposition: Geol. Mijnbouw, v. 51, p. 277-298.

Garrels, R.M. 1954. Mineral species as functions of pH and oxidation-reduction potentials, with special reference to the zone of oxidation and secondary enrichment of sulphide ore deposits: Geochim. Cosmochim. Acta, v. 5, p. 153-168.

Garrels, R.M., and C.L. Christ. 1965. Solutions, Minerals, and Equilibria. Harper and Row, New York, 450 pp.

Garrels, R.M., and F.T. Mackenzie. 1971. Evolution of Sedimentary Rocks. W.W. Norton, New York, 397 pp.

Geldsetzer, H. 1973. Syngenetic mineralization and dolomitization. In: G.C. Amstutz and A.J. Bernard, Eds., Ores in Sediments. Springer-Verlag, New York, Heidelberg, Berlin, p. 115-128.

Gibbs, R.J. 1977. Clay mineral segregation in the marine environment: Jour. Sed. Petrology, v. 42, p. 87-95.

Glasby, G.P. 1974. Mechanisms of incorporation of manganese and associated trace elements in marine manganese nodules: Oceangr. Marine Biol. Ann. Rev., v. 12, p. 11-40.

Glasby, G.P., Ed. 1977. Marine Manganese Deposits. Elsevier, Amsterdam, 523 pp.

Glasby, G.P., and A.J. Read. 1976. Deep-sea manganese nodules. In: K.H. Wolf, Ed., Handbook of Strata-Bound and Stratiform Ore Deposits. Elsevier, Amsterdam, v. 7, p. 295-340.

Goble, R.J. 1981. The leaching of copper from anilite and the production of a metastable copper sulfide structure: Canadian Min., v. 19, p. 583-591.

Godwin, C.I., and A.J. Sinclair. 1982. Average age lead isotope growth curves for shale-hosted zinc-lead deposits, Canadian Cordillera: Econ. Geol., v. 77, p. 675-690.

Godwin, C.I., A.J. Sinclair, and B.D. Ryan. 1982. Lead isotope models for the genesis of carbonate-hosted Zn-Pb, shale-hosted Ba-Zn-Pb, and silver-rich deposits in the northern Canadian Cordillera: Econ. Geol., v. 77, p. 82-94.

Goldhaber, M.B., R.L. Reynolds, and R.O. Rye. 1978. Origin of a South Texas uranium deposit: II. Sulfide petrology and sulfur isotope studies: Econ. Geol., v. 73, p. 1690-1705.

Goldsmith, J.R., and D.L. Graf. 1957. The system $CaO-MnO-CO_2$: solid solution and decomposition relations: Geochim. Cosmochim. Acta, v. 11, p. 310-334.

Gole, M.J. 1980. Mineralogy and petrology of very low metamorphic grade Archean banded iron-formations, Weld Range, Western Australia: Amer. Min., v. 65, p. 8-25.

Gole, M.J. 1981. Archean banded iron-formations, Yilgarn Block, Western Australia: Econ. Geol., v. 76, p. 1954-1974.

Gole, M.J., and C. Klein. 1981. Banded iron-formations through much of Precambrian time: Jour. Geol., v. 89, p. 169-183.

Golightly, J.P. 1979a. Geology of Soroako nickeliferous laterite deposits. *In:* D.J.I. Evans, R.S. Shoemaker, and H. Veltman, Eds., International Laterite Symposium. Amer. Inst. Mining, Metal., and Petroleum Engineers, New York, p. 38-56.

Golightly, J.P. 1979b. Nickeliferous laterites, a general description. *In:* D.J.I. Evans, R.S. Shoemaker, and H. Veltman, Eds., International Laterite Symposium. Amer. Inst. Mining, Metal., and Petroleum Engineers, New York, p. 24-37.

Golightly, J.P. 1981. Nickeliferous laterite deposits: Econ. Geol., 75 Anniv. Vol., p. 710-735.

Goodwin, A.M. 1956. Facies relationships in the Gunflint iron formation: Econ. Geol., v. 51, p. 565-595.

Goodwin, A.M. 1973a. Archean iron-formation and tectonic basins of the Canadian Shield: Econ. Geol., v. 68, p. 915-933.

Goodwin, A.M. 1973b. Archean volcanogenic iron-formation of the Canadian shield. *In:* Genesis of Precambrian Iron and Manganese Deposits. UNESCO, Paris, p. 23-33.

Goodwin, A.M., J. Monster, and H.G. Thode. 1976. Carbon and sulfur isotope abundances in Archean iron-formations and Early Precambrian life: Econ. Geol., v. 71, p. 870-891.

Gordon, M., J.I. Tracey, and M.W. Ellis. 1958. Geology of the Arkansas Bauxite Region. U.S. Geol. Survey, Prof. Paper 299, 268 pp.

Goudie, A. 1973. Duricrusts in Tropical and Subtropical Landscapes. Oxford Press, London, 174 pp.

Graf, J.L. 1978. Rare earth elements, iron formations and seawater: Geochim. Cosmochim. Acta, v. 42, p. 1845-1850.

Grandstaff, D.E. 1974. Microprobe analyses of uranium and thorium in uraninite from the Witwatersrand, South Africa, and Blind River, Ontario, Canada: Trans. Geol. Soc. South Africa, v. 77, p. 291-294.

Grandstaff, D.E. 1976. A kinetic study of the dissolution of uraninite: Econ. Geol., v. 71, p. 1493-1506.

Grandstaff, D.E. 1980. Origin of uraniferous conglomerates at Elliot Lake, Canada and in Witwatersrand, South Africa: implications for oxygen in the Precambrian atmosphere: Precambrian Res., v. 13, p. 1-26.

Grandstaff, D.E. 1981. Uraninite oxidation and the Precambrian atmosphere. In: F.C. Armstrong, Ed., Genesis of Uranium- and Gold-Bearing Precambrian Quartz-Pebble Conglomerates. U.S. Geol. Survey, Prof. Paper 1161, p. C1-C16.

Granger, H.C., and G.C. Warren. 1969. Unstable sulfur compounds and the origin of roll-type uranium deposits: Econ. Geol., v. 73, p. 160-171.

Gray, A. 1932. The Mufulira copper deposit, Northern Rhodesia: Econ. Geol., v. 27, p. 315-343.

Grew, E.S. 1974. Carbonaceous material in metamorphic rocks of New England and other areas: Jour. Geol., v. 82, p. 50-73.

Grieg, J.A., H. Baadsgaard, G.L. Cumming, R.E. Folinsbee, H.R. Krouse, H. Ohmoto, A. Sasaki, and V. Smejkal. 1971. Lead and sulphur isotopes of the Irish base metal mines in Carboniferous carbonate host rocks: Soc. Mining Geol. Japan, Spec. Issue 2, p. 84-92.

Grogan, R.M., and J.C. Bradbury. 1967. Origin of the stratiform fluorite deposits of Southern Illinois. In: J.S. Brown, Ed., Genesis of Stratiform Lead-Zinc-Barite-Fluorite Deposits in Carbonate Rocks. Economic Geology Publishing Co., Lancaster, Pa., p. 40-51.

Grønvold, F., and B.F. Westrum. 1976. Heat capacities of iron disulfides. Thermodynamics of marcasite from 5 to 700K, pyrite from 300 to 780K, and the transformation of marcasite to pyrite: Jour. Chem. Thermo., v. 8, p. 1039-1048.

Gross, G.A. 1965. Geology of iron deposits in Canada. Vol. 1. General geology and evaluation of ore deposits: Canada Geol. Survey, Econ. Geol. Rept. 22, 181 pp.

Gross, G.A. 1970. Nature and occurrence of iron ore deposits In: Survey of World Iron Ore Resources. United Nations, Publication ST/ECA/113, p. 13-31.

Gross, G.A. 1972. Primary features in cherty iron-formations: Sed. Geol., v. 7, p. 241-261.

Gross, G.A. 1980. A classification of iron formations based on depositional environments: Canadian Min., v. 18, p. 215-222.

Gross, G.A., and C.R. McLeod. 1980. A preliminary assessment of the chemical composition of iron formations in Canada: Canadian Min., v. 18, p. 223-229.

Gross, S., and L. Heller. 1963. A natural occurrence of bayerite: Min. Mag., v. 33, p. 723-724.

Gross, W.H. 1968. Evidence for a modified placer origin for auriferous conglomerates, Canavieiras miner, Jacobina, Brazil: Econ. Geol., v. 63, p. 271-276.

Grubb, P.L.C. 1970. Mineralogy, geochemistry, and genesis of the bauxite deposits on the Gove and Mitchell plateaux, Northern Australia: Min. Depos., v. 5, p. 248-272.

Grubb, P.L.C. 1971. Silicates and their paragenesis in the Brockman Iron Formation of the Wittenoom Gorge, Western Australia: Econ. Geol., v. 66, p. 281-292.

Gryaznov, V.I., and I.M. Barg. 1975. Deposition of subore sand of the Nikopol manganese basin: Dokl. Akad. Nauk, v. 220, p. 61-62.

Gulson, B.L. 1975. Differences in lead isotope composition in the stratiform McArthur zinc-lead-silver deposit: Min. Depos., v. 10, p. 277-286.

Gustafson, L.B., and N. Williams. 1981. Sediment-hosted stratiform deposits of copper, lead, and zinc: Econ. Geol., 75th Anniv. Vol., p. 139-178.

Gwosdz, W., and W. Krebs. 1977. Manganese halo surrounding Meggen ore deposit, Germany: Trans. Inst. Mining Metal., v. 86, p. B73-B77.

Gygi, R.A. 1981. Oolitic iron formations: marine or not marine?: Eclogae Geol. Helv., v. 74, p. 223-254.

Hackett, J.P., and J.L. Bischoff. 1973. New data on the stratigraphy, extent, and geologic history of the Red Sea geothermal deposits: Econ. Geol., v. 68, p. 553-564.

Hager, J. 1980. Sorption of manganese and silica by clay and carbonate: Marine Chem., v. 9, p. 199-209.

Hagni, R.D. 1976. Tri-state ore deposits: the character of their host rocks and their genesis. In: K.H. Wolfe, Ed., Handbook of Strata-Bound and Stratiform Ore Deposits. Elsevier, Amsterdam, v. 6, p. 457-491.

Hagni, R.D. 1982. The influence of original host rock character upon alteration and mineralization in the Tri-State district of Missouri, Kansas, and Oklahoma, U.S.A. In: G.C. Amstutz, A. El Goresy, G. Frenzel, C. Kluth, G. Moh, A. Wauschkuhn, and R.A. Zimmermann, Eds., Ore Genesis: The State of the Art. Springer-Verlag, New York, Heidelberg, Berlin, p. 97-107.

Hagni, R.D., and D. Gann. 1976. Microscopy of copper ore at the Creta mine, southwestern Oklahoma: Oklahoma Geol. Survey, Circular 77, p. 40-50.

Haldemann, E.G., R. Buchan, J.H. Blowes and T. Chandler. 1979. Geology of lateritic nickel deposits, Dominican Republic. In: D.J.I. Evans, R.S. Shoemaker, and H. Veltman, Eds., International Laterite Symposium. Amer. Inst. Mining, Metal., and Petroleum Engineers, New York, p. 57-84.

Hallam, A. 1963. Observations on the palaeoecology and ammonite sequence of the Frodingham Ironstone (Lower Jurassic): Paleontology, v. 6, p. 554-574.

Hallam, A. 1966. Depositional environment of British Liassic ironstones considered in the context of their facies relationships: Nature, v. 209, p. 1306-1307.

Hallam, A. 1967a. An environmental study of the Upper Domerian and Lower Toarcian in Great Britain: Phil. Trans. Royal Soc. London, v. B252, p. 393-445.

Hallam, A. 1967b. Siderite and calcite-bearing concretionary nodules in the Lias of Yorkshire: Geol. Mag., v. 104, p. 222-227.

Hallam, A. 1975. Jurassic Environments. Cambridge University Press, Cambridge, 269 pp.

Hallam, A., and M.J. Bradshaw. 1979. Bituminous shales and oolitic ironstones as indicators of transgressions and regresssions: Jour. Geol. Soc., v. 136, p. 157-164.

Hallbauer, D.K. 1981. Geochemistry and morphology of mineral components from the fossil gold and uranium placers of the Witwatersrand. In: F.C. Armstrong, Ed., Genesis of Uranium- and Gold-Bearing Precambrian Quartz-Pebble Conglomerates. U.S. Geol. Survey, Prof. Paper 1161, p. M1-M22.

Hallbauer, D.K., and T. Utter. 1977. Geochemical and morphological characteristics of gold particles from recent river deposits and the fossil placers of the Witwatersrand: Min. Depos., v. 12, p. 293-306.

Hangari, K.M., S.N. Ahmad, and E.C. Perry. 1980. Carbon and oxygen isotope ratios in diagenetic siderite and magnetite from Upper Devonian ironstone, Wadi Shatti District, Libya: Econ. Geol., v. 75, p. 538-545.

Hannak, W.W. 1982. Genesis of the Rammelsberg ore deposit near Goslar/ Upper Harz, Federal Republic of Germany, *In:* K.H. Wolfe, Ed., Handbook of Strata-Bound and Stratiform Ore Deposits. Elsevier, Amsterdam, v. 9 (in press).

Harańczyk, C. 1970. Zechstein lead-bearing shales in the fore-Sudetian monocline in Poland: Econ. Geol., v. 65, p. 481-495.

Harley, D.N. 1979. A mineralized Ordovician resurgent caldera complex in the Bathurst-Newcastle mining district, New Brunswick, Canada: Econ. Geol., v. 74, p. 786-796.

Harrell, J.A., and K.A. Eriksson. 1979. Empirical conversion equations for thin-section and sieve-derived size distributions parameters: Jour. Sed. Petrology, v. 49, p. 273-280.

Harrison, C.G.A., and M.N.A. Peterson. 1965. A magnetic mineral from the Indian Ocean: Amer. Min., v. 50, p. 704-712.

Harrison, J.E. 1972. Precambrian Belt Basin of northwestern United States: its geometry, sedimentation, and copper occurrences: Geol. Soc. Amer. Bull., v. 83, p. 1215-1240.

Harshman, E.N. 1974. Distribution of elements in some roll-type uranium deposits. *In:* Formation of Uranium Deposits. Intl. Atomic Energy Agency, Vienna, p. 169-183.

Hart, R.A. 1973. A model for chemical exchange in the basalt-seawater system of oceanic layer II: Canadian Jour. Earth Sci., v. 10, p. 799-816.

Hartmann, M. 1969. Investigation of Atlantis II Deep samples taken by FS Meteor. *In:* E.T. Degens and D. Ross, Eds., Hot Brines and Recent Heavy Metal Deposits in the Red Sea. Springer-Verlag, New York, Heidelberg, Berlin, p. 204-207.

Hathaway, J.C., and S.O. Schlanger. 1965. Nordstrandite (Al_2O_3-$3H_2O$) from Guam: Amer. Min., v. 50, p. 1029-1037.

Hayes, A.O. 1915. Wabana Iron Ore of Newfoundland. Canada Geol. Survey, Memoir 78, 162 pp.

Haymon, R.M., and M. Kastner. 1981. Hot spring deposits on the East Pacific Rise at 21°N: preliminary description of mineralogy and genesis: Earth Plan. Sci. Let., v. 53, p. 363-381.

Haynes, S.J., and M.A. Mostaghel. 1982. Present-day precipitation of lead and zinc from groundwaters: Min. Depos., v. 17, p. 213-228.

Heath, G.R. 1974. Dissolved silica and deep-sea sediments: Soc. Econ. Paleontol. Min., Spec. Pub. 20, p. 73-93.

Heath, G.R. 1981. Ferromanganese nodules of the deep-sea: Econ. Geol., 75th Anniv. Vol., p. 736-765.

Hegge, M.R., and J.C. Rowntree. 1978. Geologic setting and concepts on the origin of uranium deposits in the East Alligator River region, N.T., Australia: Econ. Geol., v. 73, p. 1420-1429.

Hékinian, R., M. Fevrier, J.L. Bischoff, P. Picot, and W.C. Shanks. 1980. Sulfide deposits from the East Pacific Rise near 21°N: Science, v. 207, p. 1433-1444.

Helgeson, H.C., J.M. Delaney, H.W. Nesbitt, and D.K. Bird. 1978. Summary and critique of the thermodynamic properites of rock-forming minerals: Amer. Jour. Sci., v. 278A, p. 1-278.

Hem, J.D. 1963. Chemical Equilibria and Rates of Manganese Oxidation. U.S. Geol. Survey, Water Supply Paper 1667A, 37 pp.

Hem, J.D. 1972. Chemical factors that influence the availability of iron and manganese in aqueous systems: Geol. Soc. Amer., Spec. Paper 140, p. 17-24.

Hem, J.D. 1981. Rates of manganese oxidation in aqueous systems: Geochim. Cosmochim. Acta, v. 45, p. 1369-1374.

Hem, J.D., C.E. Roberson, C.J. Lind, and W.L. Polzer. 1973. Chemical Interactions of Aluminum with Aqueous Silica at 25°C. U.S. Geol. Survey, Water Supply Paper 1827E, 57 pp.

Hemingway, B.S., and R.A. Robie. 1978. Revised values for the Gibbs free energy of formation of $Al(OH)^{4-}_{aq}$, diaspore, boehmite and bayerite at 298.15K and 1 bar: Geochim. Cosmochim. Acta, v. 42, p. 1533-1543.

Hemmingway, J.E. 1951. Cyclic sedimentation and the deposition of ironstones in the Yorkshire Lias: Yorkshire Geol. Soc. Proc., v. 28, p. 67-74.

Heubner, J.S. 1976. The manganese oxides—a bibliographic commentary. In: D. Rumble, Ed., Oxide Minerals. Min. Soc. Amer. Short Course Notes, v. 3, p. SH1-SH17.

Hewett, D.F. 1966. Stratified deposits of the oxides and carbonates of manganese: Econ. Geol., v. 61, p. 431-461.

Heyl, A.V., G.P. Landis, and R.E. Zartman. 1974. Isotopic evidence for the origin of Mississippi Valley-type mineral deposits: a review: Econ. Geol., v. 69, p. 992-1006.

Hiemstra, S.A. 1968. The mineralogy and petrology of the uraniferous conglomerate of the Dominion Reefs mine, Klerksdorp area: Trans. Geol. Soc. South Africa, v. 71, p. 1-100.

Hill, V.G., and A.C. Eglington. 1961. Chemical characteristics of the groundwater resources of Jamaica: Econ. Geol., v. 56, p. 533-541.

Hilpert, L. 1969. Uranium Resources of Northwest New Mexico. U.S. Geol. Survey, Prof. Paper 603, 166 pp.

Ho, C., and J.M. Coleman. 1969. Consolidation and cementation of recent sediments in the Atchafalya Basin: Geol. Soc. Amer. Bull., v. 80, p. 183-192.

Hoagland, A.D. 1976. Appalachian lead-zinc deposits. In: K.H. Wolfe, Ed., Handbook of Strata-Bound and Stratiform Ore Deposits. Elsevier, Amsterdam, v. 6, p. 495-534.

Hoefs, J. 1980. Stable Isotope Geochemistry. Springer-Verlag, New York, Heidelberg, Berlin,, 208 pp.

Hoeve, J., and T.I. I. Sibbald. 1978. On the genesis of Rabbit Lake and other unconformity-type uranium deposits in northern Saskatchewan, Canada: Econ. Geol., v. 73, p. 1450-1473.

Hoffman, P.R. 1967. Algal stromatolites: use in stratigraphic correlation and paleocurrent determination: Science, v. 157, p. 1043-1045.

Hofman, H.J. 1969. Stromatolites from the Proterozoic Animikie and Sibley Groups: Geol. Survey Canada, Paper 68-69, 77 pp.

Hofman, H.J. 1973. Stromatolites: characteristics and utility: Earth Sci. Reviews, v. 9, p. 339-373.

Holland, H.D. 1973. The oceans: a possible source of iron formations: Econ. Geol., v. 68, p. 1169-1172.

Holser, W.T., and I.R. Kaplan. 1966. Isotope geochemistry of sedimentary sulfate: Chem. Geol., v. 1, p. 93-135.

Honnorez, J., B. Honnorez-Guerstein, J. Valette, A. Wauschkuhn. 1973. Present day formation of an exhalative sulfide deposit at Vulcano (Tyrrhenian Sea), Part II: Active crystallization of fumarolic sulfides in the volcanic sediments of the Baia di Levante. In: G.C. Amstutz and A.J. Bernard, Eds., Ores in Sediments. Springer-Verlag, New York, Heidelberg, Berlin, p. 139-166.

Horn, D.R. 1972. Ferromanganese Deposits on the Ocean Floor: Lamont-Doherty Geol. Observatory, Palisades, New York, 293 pp.

Hose, H.R. 1963. Jamaica type bauxites developed on limestones: Econ. Geol., v. 58, p. 62-69.

Howard, J.H. 1977. Geochemistry of selenium: formation of ferroselite and selenium behavior in the vicinity of oxidizing sulfide and uranium deposits: Geochim. Cosmochim. Acta, v. 41, p. 1665-1678.

Hower, J., M.E. Eslinger, M.E. Hower, and E.A. Perry. 1976. Mechanism of burial metamorphism of argillaceous sediment: I. Mineralogical and chemical evidence: Geol. Soc. Amer. Bull., v. 87, p. 725-737.

Hsu, K.J. 1978a. Stratigraphy of the lacustrine sedimentation in the Black Sea: Initial Reports, Deep-sea Drilling Project, v. 42, part 2, p. 509-524.

Hsu, K.J. 1978b. Late Neogene sedimentation in the Black Sea. In: A. Matter and M.E. Tucker, Eds., Modern and Ancient Lake Sediments. Intl. Assoc. Sedimentologists Spec. Pub. No. 2, p. 127-144.

Hsu, P.H. 1977. Aluminum hydroxides and oxyhydroxides. In: J.B. Dixon and S.B. Weed, Eds. Minerals in Soil Environments. Soil Sci. Soc. Amer., Madison, Wisc., p. 99-144.

Huber, N.K., and R.M. Garrels. 1953. Relation of pH and oxidation potential to sedimentary iron mineral formation: Econ. Geol., v. 48, p. 337-357.

Hudson, J.D. 1977. Stable isotopes and limestone lithification: Jour. Geol. Soc., v. 133, p. 637-660.

Hudson, J.D. 1978. Concretions, isotopes, and the diagenetic history of the Oxford Clay (Jurassic) of central England: Sedimentology, v. 25, p. 339-370.

Huebschman, R.P. 1973. Correlation of fine carbonaceous bands across a Precambrian stagnant basin: Jour. Sed. Petrology, v. 43, p. 688-699.

Hunter, R.E. 1970. Facies of iron sedimentation in the Clinton Group. In: G.W. Fisher, Ed., Studies of Appalachian Geology, Central and Southern. Wiley-Interscience, New York, p. 101-121.

Hutchinson, R.W. 1980. Massive base metal sulfide deposits as guides to tectonic evolution. In: D.W. Strangeway, Ed., The Continental Crust and Its Mineral Deposits. Geol. Assoc. Canada, Spec. Paper 20, p. 659-684.

Hutchinson, R.W., and D.L. Searle. 1970. Stratabound pyrite deposits in Cyprus and relations to other sulfide ores: Soc. Mining Geol. Japan, Spec. Issue 3, p. 198-205.

Immega, I.P., and C. Klein. 1976. Mineralogy and petrology of some metamorphic Precambrian iron-formations in southwestern Montana: Amer. Min., v. 61, p. 1117-1144.

International Atomic Energy Agency. 1974. Formation of Uranium Deposits. Vienna, 748 pp.

Irwin, H., C. Curtis, and M.L. Coleman. 1977. Isotopic evidence for source of diagenetic carbonates formed during burial of organic-rich sediments: Nature, v. 269, p. 209-213.

Ishihara, S., Ed. 1974. Geology of Kuroko Ore Deposits. Soc. Mining Geol. Japan, Spec. Issue 6, 435 pp.

Jacobsen, J.B.E. 1975. Copper deposits in time and space: Min. Sci. Engineering, v. 7, p. 337-371.

James, H.E., and F.W. VanHouten. 1979. Miocene goethitic and chamositic oolites, northeastern Colombia: Sedimentology, v. 26, p. 125-133.

James, H.L. 1954. Sedimentary facies of iron-formation: Econ. Geol., v. 49, p. 235-293.

James, H.L. 1966. Chemistry of the Iron-Rich Sedimentary Rocks. U.S. Geol. Survey, Prof. Paper 440-W, 61 pp.

James, H.L., and R.N. Clayton. 1962. Oxygen isotope fractionation in metamorphosed iron formations of the Lake Superior region and in other iron-rich rocks: Geol. Soc. Amer. Bull., Buddington Vol., p. 217-239.

Jenks, W.F. 1975. Origins of some massive pyritic ore deposits of Western Europe: Econ. Geol., v. 70, p. 488-498.

Jensen, M.L., and A.M. Bateman. 1979. Economic Mineral Deposits (3rd ed.). John Wiley & Sons, New York, 593 pp.

Johnson, K.S. 1976. Permian copper shales of southwestern Oklahoma: Oklahoma Geol. Survey, Circ. 77, p. 3-14.

Jung, W., and G. Knitzschke. 1976. Kupferschiefer in the German Democratic Republic (GDR) with special reference to the Kupferschiefer deposit in the southeastern Harz Foreland. In: K.H. Wolfe, Ed., Handbook of Strata-Bound and Stratiform Ore Deposits. Elsevier, Amsterdam, v. 6, p. 353-406.

Kakela, P.J. 1978. Iron ore: energy, labor, and capital changes with technology: Science, v. 202, p. 1151-1157.

Kastner, M., J.B. Keene, and J.M. Gieskes. 1977. Diagenesis of siliceous oozes—I. Chemical controls on the rate of opal-A to opal-CT transformation, an experimental study: Geochim. Cosmochim. Acta, v. 41, p. 1014-1059.

Kato, T. 1961. A study of the so-called garnierite from New Caledonia: Min. Jour. (Japan), v. 3, p. 107-121.

Katsikatsos, G., N. Fytrolakis and V. Perdikatsis. 1981. Contribution to the genesis of lateritic deposits of the Upper Cretaceous transgression in Attica and Central Euboea (Greece). In: S.S. Augustithis , Ed., International Symposium on Metallogeny of

Mafic and Ultramafic Complexes. v. I: Chromites and Laterites, UNESCO, Athens, p. 279-313.

Kean, B.F., P.L. Dean, and D.F. Strong. 1981. Regional geology of the Central Volcanic Belt of Newfoundland. *In:* E.A. Swanson, D.F. Strong, and J.G. Thurlow, Eds., The Buchans Orebodies: Fifty Years of Geology and Mining. Geol. Assoc. Canada, Spec. Paper no. 22, p. 65-78.

Keller, W.D. 1979. Bauxitization of syenite and diabase illustrated in scanning electron micrographs; Econ. Geol. v. 74, p. 116-124.

Kelly, W.C. 1961. Some data bearing on the origin of Jamaican bauxite: Amer. Jour. Sci., v. 259, p. 288-294.

Kessen, K.M., M.S. Woodruff, and N.K. Grant. 1981. Gangue mineral $^{87}Sr/^{86}Sr$ ratios and the origin of Mississippi Valley-type mineralization: Econ. Geol., v. 76, p. 913-920.

Kidwell, A.L., and R.R. Bower. 1976. Mineralogy and microtextures of sulfides in the Flowerpot Shale of Oklahoma and Texas: Oklahoma Geol. Survey, Circular 77, p. 61-68.

Kimberley, M.M. 1979. Origin of oolitic iron formations: Jour. Sed. Petrology, v. 49, p. 111-132. Discussion, v. 49, p. 1351-1353; v. 50, p. 295-302; v. 50, p. 1001-1004.

Kimberley, M.M., and E. Dimroth. 1976. Basic similarity of Archean to subsequent atmospheric and hydrospheric compositions as evidenced in the distributions of sedimentary carbon, sulfur, uranium, and iron. *In:* B.F. Windley, Ed., The Early History of the Earth, Wiley-Interscience, New York, p. 579-585.

Kiskyras, D., P. Chorianopoulou, and H. Papazeti. 1978. Some remarks about the mineralogical composition of the Greek bauxites. *In:* S.S. Augustithis, Ed., 4th International Congress for the Study of Bauxites, Alumina, and Aluminum. Natl. Tech. University, Athens, p. 409-433.

Kittrick, J.A. 1969. Soil minerals in the Al_2O_3-SiO_2-H_2O system and a theory of their formation: Clays Clay Min., v. 17, p. 157-167.

Klau, W., and D.E. Large. 1980. Submarine exhalative Cu-Pb-Zn deposits—a discussion of their classification and metallogenesis: Geol. Jahrbuch, Reihe D, v. 40, p. 13-58.

Klein, C. 1966. Mineralogy and petrology of the metamorphosed Wabush Iron Formation, southwestern Labrador: Jour. Petrology, v. 7, p. 246-305.

Klein, C. 1973. Changes in mineral assemblages with metamorphism of some Precambrian iron formations: Econ. Geol., v. 68, p. 1075-1088.

Klein, C. 1974. Greenalite, stilpnomelane, minnesotaite, crocidolite, and carbonates in a very low-grade metamorphic Precambrian iron formation: Canadian Min., v. 12, p. 475-498.

Klein, C. 1978. Regional metamorphism of Proterozoic iron-formation, Labrador Trough, Canada: Amer. Min., v. 63, p. 898-912.

Klein, C., and O.P. Bricker. 1977. Some aspects of the sedimentary and diagenetic environment of Proterozoic banded iron-formation: Econ. Geol., v. 72, p. 1457-1470.

Klein, C., and R.P. Fink. 1976. Petrology of the Sokoman Iron Formation in the Howells River area, at the western edge of the Labrador Trough: Econ. Geol., v. 71, p. 453-488.

Klein, C., and M.J. Gole. 1981. Mineralogy and petrology of parts of the Marra Mamba Iron Formation, Hamersley Basin, Western Australia: Amer. Min., v. 66, p. 507-525.

Knipping, H.D. 1974. The concepts of supergene versus hypogene emplacement of uranium at Rabbit Lake, Saskatchewan, Canada In: Formation of Uranium Deposits. Intl. Atomic Energy Agency, Vienna, p. 531-548.

Knoll, A., E.S. Barghoorn, and S.M. Awramik. 1978. New microorganisms from the Aphebian Gunflint iron formation, Ontario: Jour. Paleo., v. 52, p. 976-992.

Knox, R.W. 1970. Chamosite oolites from the Winter Gill ironstone (Jurassic) of Yorkshire, England: Jour. Sed. Petrology, v. 40, p. 1216-1225.

Kostelka, L., and W.E. Petrascheck. 1967. Genesis and classification of Triassic Alpine lead-zinc deposits in the Austrian region. In: J.S. Brown, Ed., Genesis of Stratiform Lead-Zinc-Barite-Fluorite Deposits in Carbonate Rocks. Economic Geology Publishing Co., Lancaster, Pa., p. 138-146.

Kostov, I. 1968. Mineralogy. Oliver & Boyd, Edinburgh, 587 pp.

Kowalik, J., R.O. Rye, and F.J. Sawkins. 1981. Stable isotope study of the Buchans polymetallic sulphide deposits. In: E.A. Swanson, D.F. Strong, and J.G. Thurlow, Eds., The Buchans Orebodies: Fifty Years of Geology and Mining. Geol. Assoc. Canada, Spec. Paper no. 22, p. 229-254.

Kranck, S.H. 1961. A study of phase equilibria in a metamorphic iron formation: Jour. Petrology, v. 2, p. 137-184.

Krauskopf, K.B. 1957. Separation of manganese from iron in sedimentary processes: Geochim. Cosmochim. Acta, v. 12, p. 61-84.

Krauskopf, K.B. 1967. Introduction to Geochemistry. McGraw Hill, New York, 326 pp.

Kräutner, H.G. 1977. Hydrothermal-sedimentary iron ores related to submarine volcanic rises: the Teluic-Ghelar type as a carbonatic equivalent of the Lahn-Dill type. In: D.D. Klemm and H-J. Schneider, Eds., Time- and Strata-Bound Ore Deposits. Springer-Verlag, New York, Heidelberg, Berlin,, p. 232-253.

Krebs, W. 1972. Facies and development of the Meggen Reef (Devonian, West Germany): Geol. Rundschau, v. 61, p. 647-671.

Krebs, W. 1982. The geology of the Meggen ore deposit, In: K.H. Wolfe, Ed., Handbook of Strata-Bound and Stratiform Ore Deposits. Elsevier, Amsterdam, v. 9 (in press).

Kronberg, B.I., J.F. Couston, B.S. Filho, W.S. Fyfe, R.A. Nash, and D. Sugden. 1979. Minor element geochemistry of the Paragominas Bauxite, Brazil: Econ. Geol., v. 74, p. 1869-1875.

Krumbein, W.C., and R.M. Garrels. 1952. Origin and classification of chemical sediment in terms of pH and oxidation-reduction potentials: Jour. Geol., v. 60, p. 1-33.

Krumbein, W.E. 1977. Environmental Biogeochemistry and Geomicrobiology. Ann Arbor Science, Ann Arbor, Michigan, p. 839-946.

Ku, T.L. 1977. Rates of accretion. *In:* G.P. Glasby, Ed., Marine Manganese Deposits. Elsevier, Amsterdam, p. 249-267.

Kuhnel, R.A., H.J. Roorda, and J.J.S. Steensma. 1978. Distribution and partitioning of elements in nickeliferous laterites: Bull. Bureau Recherches Géologique Minières, ser. 2, sect. 2, p. 191-206.

Kyle, J.R. 1982. Geology of the Pine Point lead zinc district, *In:* K.H. Wolfe, Ed., Handbook of Strata-Bound and Stratiform Ore Deposits. Elsevier, Amsterdam, v. 9 (in press).

Lambert, I.B. 1976. The McArthur zinc-lead-silver deposit: features, metallogenesis and comparisons with some other stratiform ores. *In:* K.H. Wolfe, Ed., Handbook of Strata-Bound and Stratiform Ore Deposits. Elsevier, Amsterdam, v. 6, p. 535-585.

Lambert, I.B. 1982. Constraints on the genesis of major Australian lead-zinc-silver deposits: from Ramdohr to present. *In:* G.C. Amstutz, A. El Goresy, G. Frenzel, C. Kluth, G. Moh, A. Wauschkuhn, and R.A. Zimmermann, Eds., Ore Genesis: The State of the Art. Springer-Verlag, New York, Heidelberg, Berlin, p. 625-636.

Lambert, I.B., and T. Sato. 1974. The Kuroko and associated ore deposits of Japan: a review of their features and metallogenesis: Econ. Geol., v. 69, p. 1215-1236.

Langford, F.F. 1977. Surficial origin of North Amer. pitchblende and related deposits: Amer. Assoc. Petroleum Geol. Bull., v. 61, p. 28-42.

Langmuir, D. 1969. The Gibbs Free Energies of Substances in the System $Fe-O_2-H_2O-CO_2$ at 25°C. U.S. Geol. Survey, Prof. Paper 650-B, p. 180-184.

Langmuir, D. 1971. Particle size effect on the reaction goethite-hematite + water: Amer. Jour. Sci., v. 271, p. 147-156.

Langmuir, D. 1978. Uranium solution-mineral equilibria at low temperatures with application to sedimentary ore deposits: Geochim. Cosmochim. Acta, v. 42, p. 547-569.

Large, D.E. 1980. Geological parameters associated with sediment-hosted, submarine exhalative Pb-Zn deposits: An empirical model for mineral exploration: Geol. Jahrbuch, Reihe D, v. 40, p. 59-130.

Larter, R.C.L., A.J. Boyce, and M.J. Russell. 1981. Hydrothermal pyrite chimneys from the Ballynoe baryte deposite, Silvermines, County Tipperary, Ireland: Min. Depos., v. 16, p. 309-318.

Leclerc, J., and F. Weber. 1980. Geology and genesis of the Moanda manganese deposits, Republic of Gabon. *In:* I.M. Varentsov and Gy. Grasselly, Eds., Geology and Geochemistry of Manganese. E. Schweizerbart'sche Verlagsbuchhandlung, Stuttgart, v. 2, p. 89-112.

Lelong, F., Y. Tardy, G. Grandin, J.J. Trescases, and B. Boulange. 1976. Pedogenesis, chemical weathering and processes of formation of supergene ore deposits. *In:* K.H. Wolf, Ed., Handbook of Strata-Bound and Stratiform Ore Deposits. Elsevier, Amsterdam, v. 3, p. 93-173.

Lemoalle, J., and B. Dupont. 1973. Iron-bearing oolites and the present conditions of iron sedimentation in Lake Chad (Africa). *In:* G.C. Amstutz and A.J. Bernard, Eds., Ores in Sediments. Springer-Verlag, New York, Heidelberg, Berlin,, p. 167-178.

Lepp, H., and S.S. Goldich. 1964. Origin of Precambrian iron formations: Econ. Geol., v. 59, p. 1025-1060.

Leroy, J. 1978. The Margnac and Fanay uranium deposits of the La Crouzille district (western Massif Central, France): geologic and fluid inclusion studies: Econ. Geol., v. 73, p. 1611-1634.

Leventhal, J.S. 1980a. Comparative Geochemistry of Devonian Shale Cores from the Appalachian, Illinois, and Michigan Basins. U.S. Geol. Survey, Open-file Report 80-938, 32 pp.

Leventhal, J.S. 1980b. Organic geochemistry and uranium in Grants mineral belt: New Mexico Bureau Mines, Memoir 38, p. 75-85.

Leventhal, J.S., and M.B. Goldhaber. 1978. New data for uranium, thorium, carbon and sulfur in Devonian black shale from West Virginia, Kentucky and New York: Proc. First Eastern Gas Shales Symp., Morgantown Energy Technology Center, Morgantown, WV, MERC/SP-77/5, p. 183-219.

Lindsley, D.H. 1976. The crystal chemistry and structure of oxide minerals as exemplified by the Fe-Ti oxides. *In:* D. Rumble, Ed., Oxide Minerals. Min. Soc. Amer. Short Course Notes, v. 3, p. L1-L60.

Lippmann, F. 1977. The solubility products of complex minerals, mixed crystals, and three-layer minerals: Neues Jahr. Min., Abh., v. 119, p. 243-263.

Lockwood, R.P. 1976. Geochemistry and petrology of some Oklahoma red-bed copper occurrences: Oklahoma Geol. Survey, Circular 77, p. 61-68.

Long, D.T., and E.E. Angino. 1982. The mobilization of selected trace metals from shales by aqueous solutions: effects of temperature and ionic strength: Econ. Geol., v. 77, p. 646-652.

Ludwig, K.R., and R.L. Grauch. 1980. Coexisting coffinite and uraninite in some sandstone-host uranium ores of Wyoming: Econ. Geol., v. 75, p. 296-302.

Ludwig, K.R., M.B. Goldhaber, R.L. Reynolds, and K.R. Simmons. 1982. Uranium-lead isochron age and preliminary sulfur isotope systematics of the Felder uranium deposit, south Texas: Econ. Geol., v. 77, p. 557-563.

Lupton, J.E., G.P. Klinkhammer, W.R. Normark, R. Haymon, K.C. MacDonald, R.F. Weiss, H. Craig. 1980. Helium-3 and manganese at the 21°N East Pacific Rise hydrothermal site: Earth Plan. Sci. Let., v. 50, p. 115-127.

Lurje, A.M. 1977. Zur Herkunft des Kupfers in den Bassisschichten des Zechsteins und der Kasanstufe: Zeits. angewandte Geol., v. 23, p. 270-274.

Lusk, J. 1972. Examination of volcanic-exhalative and biogenic origins for sulfur in the stratiform massive sulfide deposits of New Brunswick: Econ. Geol., v. 67, p. 169-183.

Lutzens, H., and I. Burchardt. 1972. Metallogenetische Untersuchungen an mitteldevonischen oxidischen Eisenerzen des Elbingeröder Komplexes (Harz): Zeits. angewandte Geol., v. 18, p. 481-491.

Lydon, J.W., R.D. Lancaster, and P. Karkkainen. 1979. Genetic controls of Selwyn Basin stratiform barite/sphalerite/galena deposits: an investigation of the dominant barium mineralogy of the Tea deposit, Yukon: Geol. Survey Canada, Paper 79-1B, p. 223-229.

MacDonald, J.A., and P.L.C. Grubb. 1971. Genetic implications of shales in the Brockman iron formation from Mount Tom Price and Wittenoom Gorge, Western Australia: Jour. Geol. Soc. Australia, v. 18, p. 81-86.

MacGeehan, P. 1972. Vertical zonation within the Aurukun bauxite deposit, North Queensland, Australia: 24th Intl. Geol. Cong., Sect. 4, p. 424-434.

Mackenzie, F.T., and R. Wollast. 1977. Thermodynamic and kinetic controls of global chemical cycles of the elements. *In:* W. Stumm, Ed., Global Chemical Cycles and Their Alteration by Man. Dahlem Konferenzen, Berlin, p. 45-59.

Mackenzie, F.T., and J.D. Pigott. 1981. Tectonic controls of Phanerozoic sedimentary rock cycling: Jour. Geol. Soc., v. 138, p. 183-196.

Macquar, J. Cl., and Ph. Lagny. 1981. Minéralisations Pb-Zn "sous inconformité" des series de plates-formes carbonatées. Example du gisement de Trèves (Gard, France). Relations dolomitisations, dissolutions et minéralisations: Min. Depos., v. 16, p. 283-307.

MacQueen, R.W., and R.I. Thompson. 1978. Carbonate-hosted lead-zinc occurrences in northeastern British Columbia with emphasis on the Robb Lake deposit: Canadian Jour. Earth Sci., v. 15, p. 1737-1762.

Magaritz, M., E. Gavish, and N. Baker. 1979. Carbon and oxygen isotope composition; indicators of cementation environment in Recent, Holocene, and Pleistocene sediments along the coast of Isreal: Jour. Sed. Petrology, v. 49, p. 401-411.

Majumder, T., and K.L. Chakraborty. 1977. Primary sedimentary structures in the banded iron-formation of Orissa, India: Sed. Geol., v. 19, p. 297-300.

Maksimovic, Z. 1978. Nickel in karstic environment: in bauxites and in karstic nickel deposits: Bull. Bureau Recherches Géologique Minières, ser. 2, sect. 2, p. 173-183.

Manheim, F.T. 1974. Red Sea geochemistry; Initial Reports Deep-sea Drilling Project, v. 23, p. 975-998.

Manheim, F.T., and K.M. Chan. 1974. Interstitial waters of Black Sea sediments: new data and review. *In:* E.T. Degens and D.A. Ross, Eds., The Black Sea—Geology, Chemistry, and Biology. Amer. Assoc. Petroleum Geol., Memoir No. 20, p. 155-180.

Mann, A.W., and R.L. Deutscher. 1978. Genesis principles for the precipitation of carnotite in calcrete drainages in Western Australia: Econ. Geol., v. 73, p. 1724-1737.

Markun, C.D., and A.F. Randazzo. 1980. Sedimentary structures in the Gunflint Iron Formation, Schreiber Beach, Ontario: Precambrian Res., v. 12, p. 287-310.

Marowsky, G. 1969. Schwefel-, Kohlenstoff-, und Sauerstoff isotopen untersuchungen am Kupferschiefer als Beitrag zur genetischen Deutung: Contrib. Min. Petrology, v. 22, p. 290-234.

Marsden, R.W., J.W. Emanuelson, J.S. Owens, N.E. Walker, and W.F. Werner. 1968. The Mesabi iron range, Minnesota. *In:* J.D. Ridge, Ed., Ore Deposits of the United States, Amer. Inst. Mining Metal. Petroleum Engineers, New York, v. 1, p. 518-537.

Martinsson, A. 1974. The Cambrian of Norden. *In:* C.H. Holland, Ed., Cambrian of the British Isles, Norden, and Spitsbergen, John Wiley & Sons, London, p. 185-284.

Maucher, A., and H-J. Schneider. 1967. The Alpine lead-zinc ores. *In:* J.S. Brown, Ed., Genesis of Stratiform Lead-Zinc-Barite-Fluorite Deposits in Carbonate Rocks. Economic Geology Publishing Co., Lancaster, Pa., p. 71-89.

Maynard, J.B. 1976. The long-term buffering of the oceans: Geochim. Cosmochim. Acta, v. 40, p. 1523-1532.

Maynard, J.B. 1980. Sulfur isotopes of iron sulfides from the Devonian-Mississippian shale sequence of the Appalachian Basin: control by rate of sedimentation: Amer. Jour. Sci., v. 280, p. 772-786.

Maynard, J.B. 1981. Carbon isotopes as indicators of dispersal patterns in Devonian-Mississippian shales of the Appalachian Basin: Geology, v. 9, p. 262-265.

Maynard, J.B., and S.K. Lauffenburger. 1978. A marcasite layer in prodelta turbidites of the Borden Formation (Mississipian) in Eastern Kentucky: Southeastern Geology, v. 20, p. 47-58.

McDuff, R.E., and J.M. Edmond. 1982. On the fate of sulfate during hydrothermal circulation at mid-ocean ridges: Earth Plan. Sci. Let., v. 57, p. 117-132.

McKelvey, V.E., N.A. Wright, and R.W. Rowland. 1979. Manganese nodule resources in the Northeastern Equatorial Pacific In: J.L. Bischoff and D.Z. Piper, Eds. Marine Geology and Oceanography of the Pacific Manganese Nodule Province, Plenum Press, New York, p. 747-762.

McMillan, R.H. 1977. Uranium in Canada: Canadian Petroleum Geol. Bull., v. 25, p. 1222-1249.

Mellon, G.B. 1962. Petrology of Upper Cretaceous oolitic iron-rich rocks from northern Alberta: Econ. Geol., v. 57, p. 921-940.

Mengel, J.T. 1973. Physical sedimentation in Precambrian cherty iron formation of the Lake Superior type. In: G.L. Amstutz and A.J. Bernard, Eds., Ores in Sediments. Springer-Verlag, New York, Heidelberg, Berlin,, p. 179-194.

Mindszenty, A. 1978. Tentative interpretation of the micromorphology of bauxitic laterites. In: S.S. Augustithis, Ed., 4th International Congress for the Study of Bauxites, Alumina, and Aluminum. Greece, Natl. Tech. University Athens, p. 599-613.

Minter, W.E.L. 1976. Detrital gold, uranium and pyrite concentrations related to sedimentology in the Precambrian Vaal Reef placer, Witwatersrand, South Africa: Econ. Geol., v. 71, p. 157-176.

Minter, W.E.L. 1981. The crossbedded nature of Proterozoic Witwatersrand placers in distal environments and a paleocurrent analysis of the Vaal Reef placer. In: F.C. Armstrong, Ed., Genesis of Uranium- and Gold-Bearing Precambrian Quartz-Pebble Conglomerates. U.S. Geol. Survey, Prof. Paper 1161, p. G1-G9.

Moeskops, P.G. 1977. Yilgarn nickel gossan geochemistry—a review, with new data: Jour. Geochem. Exploration, v. 8, p. 247-258.

Morey, G.B. 1972. Mesabi range. In: P.K. Sims and G.B. Morey, Eds., Geology of Minnesota, Minnesota Geol. Survey, St. Paul, p. 204-217.

Morganti, J.M. 1981. Ore deposit models—4. Sedimentary-type stratiform ore deposits: some models and a new classification: Geoscience Canada, v. 8., p. 65-75.

Morris, W.A. 1977. Paleolatitudes of glaciogenic upper Precambrian Rapitan Group and the use of tillites as chronostratigraphic marker horizons: Geology, v. 5, p. 85-88. Discussion, v. 6, p. 4-5.

Morris, R.C. 1980. A textural and mineralogical study of the relationship of iron ore to banded iron-formation in the Hamersley iron province of Western Australia: Econ. Geol., v. 75, p. 184-209.

Mottl, M.J., and H.D. Holland. 1978. Chemical exchange during hydrothermal alteration of basalt by seawater—I. experimental results for major and minor components of seawater: Geochim. Cosmochim. Acta, v. 42, p. 1103-1117.

Mottl, M.J., H.D. Holland, and R.F. Corr. 1979. Chemical exchange during hydrothermal alteration of basalt by seawater— II. Experimental results for Fe, Mn, and sulfur species: Geochim. Cosmochim. Acta, v. 45, p. 868-884.

Mposkos, E. 1981. The Ni-Fe laterite ores of Almopia zone. *In:* S.S. Augustithis , Ed., International Symposium on Metallogeny of Mafic and Ultramafic Complexes. v. I: Chromites and Laterites, UNESCO, Athens, p. 317-337.

Müller, G., and U. Förstner. 1973. Recent iron ore formation in Lake Malawi, Africa: Min. Depos., v. 8, p. 278-290.

Munday, R.J. 1978. Uranium mineralization in northern Saskatchewan: Canadian Mining Metal. Bull., v. 71, p. 1-76.

Murata, K.J., I. Friedman, and J.D. Gleason. 1977. Oxygen isotope relations between diagenetic silica minerals in Monterrey Shale, Temblor Range, California: Amer. Jour. Sci., v. 277, p. 259-272.

Murray, J.W. 1978. Gamma-FeOOH in marine sediments: Amer. Geophys. Union Trans., v. 59, p. 411-412.

Murray, J.W., V. Grundmanis, and W.M. Smethie. 1978. Interstitial water chemistry in sediments of Saanich Inlet: Geochim. Cosmochim. Acta, v. 42, p. 1011-1026.

Murray, W.J. 1975. McArthur River H.Y.C. lead-zinc and related deposits, N.T. *In:* C.L. Knight, Ed., Economic Geology of Australia and Papua New Guinea. Australasian Inst. Mining Metal., Parkville, Australia, v. 1, p. 329-339.

Nahon, D., A.V. Carozzi, and C. Parron. 1980. Lateritic weathering as a mechanism for the generation of ferruginous ooids: Jour. Sed. Petrology, v. 50, p. 1287-1298.

Nash, J.T., H.C. Granger, and S.S. Adams. 1981. Geology and concepts of genesis of important types of uranium deposits: Econ. Geol., 75th Anniv. Vol., p. 63-116.

Nelson, B.W., and R. Roy. 1958. Synthesis of the chlorites and their structural and chemical constitution: Amer. Min., v. 43, p. 707-725.

Nickel, E.H., P.D. Allchurch, M.G. Mason, and J.R. Wilmshurst. 1977. Supergene alteration at the Perseverance nickel deposit, Agnew, Western Australia: Econ. Geol., v. 72, p. 184-203.

Nickel, E.H., J.R. Ross, and M.R. Thornber. 1974. The supergene alteration of pyrrhotite-pentlandite ore at Kambalda, Western Australia: Econ. Geol., v. 69, p. 93-107.

Nicolas, J. 1968. Nouvelles donées sur la genèse des bauxites à mur karstique du sud-est de la France: Min. Depos., v. 3, p. 18-33.

Nicolini, P. 1967. Remarques comparatives sur quelques éléments sédimentologiques et páleogéographiques liés aux gisements de fer oolithiques du Djebel Ank (Tunisie) et de Lorraine (France): Min. Depos., v. 2, p. 95-101.

Nikitina, A.P., and B.B. Zvyagin. 1973. Origin and crystal structure features of clay minerals from the lateritic bauxites in the European part of the U.S.S.R.. *In:* J.M. Serratosa, Ed., Proceedings of the International Clay Conference 1972. Division de Ciencias, C.S.I.C., Madrid, p. 227-233.

Northrop, H.R., M.B. Goldhaber, G.P. Landis, R.O. Rye, and C.G. Whitney. 1982. Localization of tabular, sediment-hosted uranium-vanadium deposits of Henry structural basin, Utah: Amer. Assoc. Petroleum Geol. Bull., v. 66, p. 613.

Norton, S.A. 1973. Laterite and bauxite formation: Econ. Geol. v. 68, p. 353-361.

Oberc, J., and J. Serkies. 1968. Evolution of the fore-Sudetian copper deposit: Econ. Geol., v. 63, p. 372-379.

Ogura, Y. 1977. Mineralogical studies on the occurrence of nickeliferous laterite deposits in the Southwestern Pacific area: Mining Geol., v. 27, p. 379-399.

Ohmoto, H., and R.O. Rye. 1974. Hydrogen and oxygen isotopic compositions of fluid inclusions in the Kuroko deposits, Japan: Econ. Geol., v. 69, p. 947-953.

Oshimi, T., T. Hashimoto, H. Kamono, S. Kawabe, K. Suga, S. Tanimura, and Y. Ishikawa. 1974. Geology of the Kosaka Mine, Akita Prefecture. In: S. Ishihara, Ed., Geology of Kuroko Ore Deposits Soc. Mining Geol. Japan, Spec. Issue 6, p. 89-100.

Ostwald, J. 1975. Mineralogy of manganese oxides from Groote Eylandt: Min. Depos., v. 10, p. 1-12.

Ostwald, J. 1980. Aspects of the mineralogy, petrology, and genesis of the Groote Eylandt manganese ores. In: I.M. Varentsov and Gy. Grasselly, Eds., Geology and Geochemistry of Manganese. E. Schweizerbart'sche Verlagsbuchhandlung, Stuttgart, v. 2, p. 149-182.

Oudin, E., P. Picot, and G. Pouit. 1981. Comparison of sulphide deposits from the East Pacific Rise and Cyprus: Nature, v. 291, p. 404-407.

Pačes, T. 1978. Reversible control of aqueous aluminum and silica during the irreversible evolution of natural waters: Geochim. Cosmochim. Acta, v. 42, p. 1487-1493.

Parham, W.E. 1966. Lateral variations of clay mineral assemblages in modern and ancient sediments. In: K. Gekker and A. Weiss, Eds., Proceedings of the International Clay Conference. v. 1, p. 136-145.

Park, C.F., and R.A. MacDiarmid. 1975. Ore Deposits. W.H. Freeman and Co., San Francisco, 529 pp.

Park, D.F. 1959. The origin of hard hematite in itabirite: Econ. Geol., v. 54, p. 573-587.

Patterson, S.H., and C.E. Roberson. 1961. Weathered Basalt in the Eastern Part of Kauai, Hawaii. U.S. Geol. Survey, Prof. Paper 424C, p. 195-198.

Pawłowska, J., and H. Wedow. 1980. Strata-bound zinc-lead deposits of the Upper Silesian region, Poland—a review of some recent research. In: J.D. Ridge, Ed., Proceedings of the Fifth Quadrennial IAGOD Symposium. E. Schweizerbart'sche Verlagsbuchhandlung, Stuttgart, p. 467-486.

Pearce, J.A., T. Alabaster, A.W. Shelton, and M.P. Searle. 1981. The Oman ophiolite as a Cretaceous arc-basin complex: evidence and implications: Phil. Trans. Royal Soc. London, v. A300, p. 299-317.

Pedersen, T.F., and N.B. Price. 1982. The geochemistry of manganese carbonate in Panama Basin sediment: Geochim. Cosmochim. Acta, v. 46, p. 59-68.

Pedro, G., J.P. Carmouze, and B. Velde. 1978. Peloidal nontronite formation in recent sediments in Lake Chad: Chem. Geol., v. 23, p. 139-149.

Pequegnat, W.E., W.R. Bryant, A.D. Fredericks, T. McKee, and R. Spalding. 1972. Deep-sea ironstone deposits in the Gulf of Mexico: Jour. Sed. Petrology, v. 42, p. 700-710.

Perkins, D., E.J. Essene, E.F. Westrum, and V.J. Wall. 1979. New thermodynamic data for diaspore and their application to the system Al_2O_3-SiO_2-$H2O$: Amer. Min., v. 64, p. 1080-1090.

Perry, E.C., and S.N. Ahmad. 1981. Oxygen and carbon isotope geochemistry of the Krivoy Rog iron formation, Ukranian SSR: Lithos, v. 14, p. 83-92.

Perry, E.C., F.C. Tan, and G.B. Morey. 1973. Geology and stable isotope geochemistry of the Biwabik Iron Formation, Northeastern Minnesota: Econ. Geol., v. 68, p. 1110-1125.

Petersen, U. 1971. Laterite and bauxite formation: Econ. Geol. v. 66, p. 1070-1071.

Peterson, N.P. 1954. Copper cities copper deposit, Globe-Miami district, Arizona: Econ. Geol., v. 49, p. 362-377.

Petruk, W. 1977. Mineralogical characteristics of an oolitic iron deposit in the Peace River District, Alberta: Canadian Min., v. 15, p. 3-13.

Petruk, W., D.M. Farrell, E.E. Laufer, R.J. Tremblay, and P.G. Manning. 1977. Nontronite and ferruginous opal from the Peace River iron deposit in Alberta, Canada: Canadian Min., v. 15, p. 14-21. Discussion, v. 16, p. 119.

Piper, D.Z., and M.E. Williamson. 1977. Composition of Pacific Ocean ferromanganese nodules: Marine Geol., v. 23, p. 285-303.

Porath, H. 1967. Paleomagnetism and the age of Australian hematite ore bodies: Earth Plan. Sci. Let., v. 2, p. 409-414.

Porrenga, D.H. 1965. Chamosite in recent sediments of the Niger and Orinoco Deltas: Geol. Mijnbouw, v. 44, p. 400-403.

Porrenga, D.H. 1967. Glauconite and chamosite as depth indicators in the marine environment: Marine Geology, v. 5, p. 495-501.

Potter, P.E., J.B. Maynard, and W.A. Pryor. 1980. Sedimentology of Shale. Springer-Verlag, New York, Heidelberg, Berlin, 306 pp.

Potter, P.E., J.B. Maynard, and W.A. Pryor. 1982. Gas-bearing Devonian shales of the Appalachian basin, statements and discussion: Oil & Gas Jour., Jan 25, p.290-316.

Potter, R.W. 1977. An electrochemical investigation of the system copper-sulfur: Econ. Geol., v. 72, p. 1524-1542.

Pouit, G. 1976. La concentration de manganèse de l'Imini (Maroc), peut-elle être d'origine karstique?: Compte Rendu Soc. géol. France, No. 5, p. 227-229.

Pouit, G. 1978. Différents modèles de minéralisations "hydrothermal sédimentaire" a Zn (Pb) du Paléozoique des Pyrenees centrales: Min. Depos., v. 13, p. 411-421.

Pretorius, D.A. 1975. The depositional environment of the Witwatersrand goldfields: a chronological review of speculations and observations: Min. Sci. Engineering, v. 7, p. 18-47.

Pretorius, D.A. 1976. The nature of the Witwatersrand gold-uranium deposits. In: K.H. Wolf, Ed., Handbook of Strata-Bound and Stratiform Ore Deposits. Elsevier, Amsterdam, v. 7, p. 29-88.

Pretorius, D.A. 1981. Gold and uranium in quartz-pebble conglomerates: Econ. Geol., 75th Anniv. Vol., p. 117-138.

Price, N.B., and S.E. Calvert. 1970. Compositional variation in Pacific Ocean ferro-manganese nodules and its relationship to sediment accumulation rates: Marine Geol., v. 9, p. 145-171.

Provo, L.J. 1977. Stratigraphy and Sedimentology of Radioactive Devonian-Mississippian Shales of the Central Appalachian Basin. Unpubl. PhD Dissertation, University of Cincinnati, 207 pp.

Provo, L.J., R.C. Kepferle, P.E. Potter. 1978. Division of black Ohio Shale in eastern Kentucky: Amer. Assoc. Petroleum Geol., v. 62, p. 1703-1713.

Puchelt, H. 1973. Recent iron sediment formation at the Kameni Islands, Santorini (Greece). In: G.C. Amstutz and A.J. Bernard, Eds., Ores in Sediments. Springer-Verlag, New York, Heidelberg, Berlin, p. 227-246.

Puchelt, H., H.H. Schock, E. Schroll. 1973. Rezente marine Eisenerze auf Santorin, Greichenland: I. Geochemie, Entstehung, Mineralogie: Geol. Rundschau, v. 62, p. 786-803.

Pulfrey, W. 1933. The iron-ore oolites and pisolites of North Wales: Quart. Jour. Geol. Soc. London, v. 89, p. 401-430.

Quade, H. 1976. Genetic problems and environmental features of volcano-sedimentary iron-ore deposits of the Lahn-Dill type. In: K.H. Wolfe, Ed., Handbook of Strata-Bound and Stratiform Ore Deposits. Elsevier, Amsterdam, v. 7, p. 255-294.

Rackley, R.I. 1976. Origin of western-states type uranium deposition. In: K.H. Wolf, Ed., Handbook of Strata-Bound and Stratiform Ore Deposits. Elsevier, Amsterdam, p. 89-156.

Rai, K.L., S.N. Sarkar, and P.R. Paul. 1980. Primary depositional and diagenetic features in the banded iron formation and associated iron-ore deposits of Noamundi, Singh-bhum District, Bihar, India: Min. Depos., v. 15, p. 189-200.

Ranger, M.R. 1979. The Sedimentology of a Lower Paleozoic Peritidal Sequence and Associated Iron Formations, Bell Island, Conception Bay, Newfoundland. Unpubl. MSc Thesis, Memorial University of Newfoundland, 125 pp.

Renfro, A.R. 1974. Genesis of evaporite-associated stratiform metalliferous deposits: a sabhka process: Econ. Geol., v. 69, p. 33-45. Discussion, v. 70, p. 407-409.

Rentzsch, J. 1974. The Kupferschiefer in comparison with the the deposits of the Zambian copperbelt. In: P. Bartholomé, Ed., Gisements Stratiformes et Provinces Cuprifères, Soc. géol. Belgique, Liège, p. 403-426.

Reynolds, R.L., and M.B. Goldhaber. 1978. Origin of a South Texas uranium deposit: I. Alteration of iron-titanium oxide minerals: Econ. Geol., v. 73, p. 1677-1689.

Reynolds, R.L., M.B. Goldhaber, and D.J. Carpenter. 1982. Biogenic and nonbiogenic ore-forming processes in the south Texas uranium district: evidence from the Panna Maria deposit: Econ. Geol., v. 77, p. 541-556.

Rich, J.L. 1951. Probable fondo origin of Marcellus-Ohio-New Albany-Chattanooga bituminous shales: Amer. Assoc. Petroleum Geol. Bull., v. 35, p. 2017-2040.

Rich, R.A., H.D. Holland, and U. Petersen. 1977. Hydrothermal Uranium Deposits. Elsevier, Amsterdam, 264 pp.

Rickard, D.T. 1969a. The microbiological formation of iron sulfides: Stockholm Contributions Geol., v. 20, p. 49-66.

Rickard, D.T. 1969b. The chemistry of iron sulfide formation at low temperature: Stockholm Contributions Geol., v. 20, p. 67-95.

Rickard, D.T. 1972. Covellite formation in low temperature aqueous solutions: Min. Depos., v. 7, p. 180-188.

Rickard, D.T. 1973. Limiting conditions for synsedimentary sulfide ore formation: Econ. Geol., v. 68, p. 605-617.

Rickard, D.T., M.Y. Willden, N-E. Marinder, and T.H. Donnelly. 1979. Studies on the genesis of the Laisvall Sandstone lead-zinc deposit, Sweden: Econ. Geol., v. 74, p. 1255-1285. Discussion, v. 76, p. 2047-2060.

Ripley, E.M., M.W. Lambert, and P. Berendsen. 1980. Mineralogy and paragenesis of red-bed copper mineralization in the Lower Permian of south central Kansas: Econ. Geol., v. 75, p. 722-729.

Rivers, T., and R. Wardle. 1979. Labrador trough: 2.3 billion years of history: Newfoundland Jour. Geol. Education, v. 4, p. 19-44.

Robertson, A.H.F. 1975. Cyprus umbers: basalt-sediment relationsips on a Mesozoic ocean ridge: Jour. Geol. Soc., v. 131, p. 511-531.

Robertson, A.H.F. 1978. Metallogenesis along a fossil fracture zone: Arakapas fault belt, Troodos Massif, Cyprus: Earth Plan. Sci. Let., v. 41, p. 317-329.

Robertson, A.H.F., and J.D. Hudson. 1973. Cyprus umbers: chemical precipitates on a Tethyan ocean ridge: Earth Plan. Sci. Let., v. 18, p. 93-101.

Robertson, D.S. 1974. Proterozoic units as fossil time markers and their use in uranium prospection. In: Formation of Uranium Ore Deposits. Intl. Atomic Energy Agency, Vienna, p. 495-512.

Robertson, D.S., J.E. Tilsley and G.M. Hogg. 1978. The time-bound character of uranium deposits: Econ. Geol., v. 73, p. 1409-1419.

Robertson, J.A. 1973. A review of recently acquired geologic data, Blind River-Elliot Lake area: Geol. Assoc. Canada Spec. Pap. 12, p. 169-198.

Robie, R.S., B.S. Hemingway, and J.R. Fisher. 1978. Thermodynamic properties of minerals and related substances at 298.15K and 1 Bar (10^5 Pascals) Pressure and at Higher Temperatures. U.S. Geol. Survey, Bull. 1452, 456 pp.

Roch, E. 1966. A comparison of some European bauxites with those of the Caribbean: Jour. Geol. Soc. Jamaica, v. 8, p. 1-23.

Rodgers, J. 1945. Manganese content of the Shady Dolomite in Bumpass Cove, Tennessee: Econ. Geol., v. 40, p. 129-135.

Roedder, E. 1976. Fluid-inclusion evidence on the genesis of ores in sedimentary and volcanic rocks. In: K.H. Wolfe, Ed., Handbook of Strata-Bound and Stratiform Ore Deposits. Elsevier, Amsterdam, v. 2, p. 67-134.

Roedder, E. 1977. Fluid inclusion studies of ore deposits in the Viburnum Trend, Southeast Missouri: Econ. Geol., v. 72, p. 474-479.

Rohrlich, V. 1974. Microstructure and microchemistry of iron ooliths: Min. Depos., v. 9, p. 133-142.

Rohrlich, V., N.B. Price, and S.E. Calvert. 1969. Chamosite in the Recent sediments of Loch Etive, Scotland: Jour. Sed. Petrology, v. 39, p. 624-631.

Roscoe, S.W. 1973. The Huronian supergroup, a paleo-Aphebian succession showing evidence of atmospheric evolution: Geol. Assoc. Canada, Spec. Pap. 12, p. 31-47.

Rose, A.W. 1976. The effect of cuprous chloride complexes in the origin of red-bed and related deposits: Econ. Geol., v. 71, p. 1036-1048.

Rosholt, J.N., Prijana, and D.C. Noble. 1971. Mobility of uranium and thorium in glassy and crystallized silicic volcanic rocks: Econ. Geol., v. 66, p. 1061-1069.

Rouse, J.E., and N. Sherif. 1980. Major evaporite deposition from groundwater-remobilized salts: Nature, v. 285, p. 470-472.

Roy, S. 1976. Ancient manganese deposits. *In:* K.H. Wolf, Ed., Handbook of Strata-Bound and Stratiform Ore Deposits. Elsevier, Amsterdam, v. 7, p. 395-476.

Roy, S. 1981. Manganese Deposits. Academic Press, London, 458 pp.

Russell, M.J. 1975. Lithogeochemical environment of the Tynagh base-metal deposit, Ireland, and its bearing on ore deposition: Trans. Inst. Mining Metal., v. 84, p. B128-B133.

Rye, D.M., and N. Williams. 1981. Studies of the base metal sulfide deposits at McArthur River, Northern Territory, Australia: III. The stable isotope geochemistry of the H.Y.C., Ridge, and Cooley deposits: Econ. Geol., v. 76, p. 1-26. Discussion, v. 76, p. 2257-2260.

Rye, R.O., and H. Ohmoto. 1974. Sulfur and carbon isotopes and ore genesis: a review: Econ. Geol., v. 69, p. 826-842.

Saager, R. 1981. Geochemical studies on the origin of the detrital pyrites in the conglomerates of the Witwatersrand goldfields, South Africa. *In:* F.C. Armstrong, Ed., Genesis of Uranium- and Gold-Bearing Precambrian Quartz-Pebble Conglomerates. U.S. Geol. Survey, Prof. Paper 1161, p. L1-L17.

Saager, R., M. Meyer, and R. Muff. 1982a. Gold distribution in supracrustal rocks from Archean greenstone belts of southern Africa and from Paleozoic ultramafic complexes of the European Alps: metallogenic and geochemical implications: Econ. Geol., v. 77, p. 1-27.

Saager, R., T. Utter, and M. Meyer. 1982b. Pre-Witwatersrand and Witwatersrand conglomerates in South Africa: a mineralogical comparison and bearings on the genesis of gold-uranium placers. *In:* G.C. Amstutz, A. El Goresy, G. Frenzel, C. Kluth, G. Moh, A. Wauschkuhn, and R.A. Zimmermann, Eds., Ore Genesis: The State of the Art. Springer-Verlag, New York, Heidelberg, Berlin, p. 38-56.

Sahasrabudhe, Y.S. 1978. Geochemistry of bauxite profiles on different rock types from central and western India. *In:* S.S. Augustithis, Ed., 4th International Congress for the study of Bauxites, Alumina, and Aluminum. Greece, Natl. Tech. University Athens, p. 734-751.

Sakai, H., and O. Matsubaya. 1971. Sulfur and oxygen isotopic ratios of gypsum and barite in the black ore deposits of Japan: Soc. Mining Geol. Japan, Spec. Issue 2, p. 80-83.

Sakai, H., and O. Matsubaya. 1974. Isotopic geochemistry of the thermal waters of Japan and its bearing on the Kuroko ore solutions: Econ. Geol., v. 69, p. 974-991.

Sakai, V., E. Gunnlaugsson, J. Tomasson, and J.E. Rouse. 1980. Sulfur isotope systematics in Icelandic geothermal systems and influence of seawater circulation at Reykjanes: Geochim. Cosmochim. Acta, v. 44, p. 1223-1231.

Sanford, R.F. 1982. Preliminary model of regional Mesozoic groundwater flow and uranium deposition in the Colorado Plateau: Geology, v. 10, p. 348-352.

Sangster, D.F. 1968. Relative sulphur isotope abundances of ancient seas and stratabound sulphide deposits: Proc. Geol. Assn. Canada, v. 19, p. 79-81.

Sangster, D.F. 1976. Carbonate-hosted lead-zinc deposits. *In:* K.H. Wolfe, Ed., Handbook of Strata-Bound and Stratiform Ore Deposits. Elsevier, Amsterdam, v. 6, p. 447-454.

Sato, M. 1960. Oxidation of sulfide ore bodies II. Oxidation mechanism of sulfide minerals at 25°C: Econ. Geol., v. 55, p. 1202-1231.

Sato, T. 1972. Behaviours of ore-forming solutions in seawater: Mining Geol., v. 22, p. 31-42.

Sawkins, F.J. 1976a. Massive sulphide deposits in relation to geotectonics. *In:* D.F. Strong, Ed., Metallogeny and Plate Tectonics. Geol. Assoc. Canada, Spec. Publ. 14, p. 221-242.

Sawkins, F.J. 1976b. Metal deposits related to intracontinental hotspots and rifting environments: Jour. Geol., v. 84, p. 653-671.

Schellmann, W. 1969. Die Bildungsbedingungen sedimentärer Chamosit- und Hämatit-Eisenerze am Beispiel der Lagerstätte Echte: Neues Jahr. Min., Abh., v. 111, p. 1-31.

Schellmann, W. 1971. Über Beziehungen lateritischer Eisen-, Nickel-, Aluminun-, und Manganerze zu ihren Ausgangsteinen: Min. Depos., v. 6, p. 275-291.

Schellmann, W. 1978. Behaviour of nickel, cobalt and chromium in ferruginous lateritic nickel ores: Bull. Bureau Recherches Géologique Minières, ser. 2, sec. 2, p. 275-282.

Schellmann, W. 1982. Formation of nickel silicate ores by weathering of ultramafic rocks. *In:* H. Van Olphen and F. Veniale, Eds., International Clay Conference 1981, Elsevier, Amsterdam, p. 623-634.

Schermerhorn, L.J.G. 1971. Pyrite emplacement by gravity flow: Boletin Geologico y Minero, v. 82, p. 304-308.

Schermerhorn, L.J.G. 1974. Late Precambrian mixtites: glacial and/or nonglacial?: Amer. Jour. Sci., v. 274, p. 673-824.

Schidlowski, M. 1968. The gold fraction of the Witwatersrand conglomerates from the Orange Free State goldfield (South Africa): Min. Depos., v. 3, p. 344-363.

Schidlowski, M. 1976. Archean atmosphere and evolution of the terrestrial oxygen budget. *In:* B.F. Windley, Ed., The Early History of the Earth, Wiley-Interscience, New York, p. 525-535.

Schidlowski, M. 1981. Uraniferous constituents of the Witwatersrand conglomerates: ore microscopic observations and implications for the Witwatersrand metallogeny. *In:* F.C. Armstrong, Ed., Genesis of Uranium- and Gold-Bearing Precambrian Quartz-Pebble Conglomerates. U.S. Geol. Survey, Prof. Paper 1161, p. N1-N29.

Schidlowski, M., C.E. Junge, and H. Pietrek. 1977. Sulfur isotope variations in marine sulfate evaporites and the Phanerozoic oxygen budget: Jour. Geophys. Res., v. 82, p. 2557-2565.

Schneider, H-J. 1964. Facies differentiation and controlling factors for the depositional lead-zinc concentration in the Ladinian geosyncline of the eastern Alps. *In:* G.C. Amstutz, Ed., Sedimentology and Ore Genesis. Elsevier, Amsterdam, p. 29-46.

Schneider, H-J., P. Moller, P.P. Parekh, and E. Zimmer. 1977. Fluorine contents in carbonate sequences and rare earths distribution in fluorites of Pb-Zn deposits in east-alpine mid-Triassic: Min. Depos., v. 12, p. 22-36. Discussion, v. 13, p. 281-287.

Schoen, R. 1964. Clay minerals in the Silurian Clinton ironstones, New York State: Jour. Sed. Petrology, v. 34, p. 855-863.

Schoen, R., and C.E. Roberson. 1970. Structure of aluminum hydroxide and geochemical implications: Amer. Min., v. 55, p. 43-77.

Schopf, J.W. 1981. The Precambrian development of an oxygenic atmosphere. *In:* F.C. Armstrong, Ed., Genesis of Uranium- and Gold-Bearing Precambrian Quartz-Pebble Conglomerates. U.S. Geol. Survey, Prof. Paper 1161, p. B1-B11.

Schultz, R.W. 1966. Lower Carboniferous cherty ironstones at Tynagh, Ireland: Econ. Geol., v. 61, p. 311-342.

Schulz, O. 1964. Lead-zinc deposits in the Calcareous Alps as an example of submarine-hydrothermal formation of mineral deposits. *In:* G.C. Amstutz, Ed., Sedimentology and Ore Genesis. Elsevier, Amsterdam, p. 47-52.

Schulz, O. 1982. Karst or thermal mineralizations interpreted in the light of sedimentary ore fabrics. *In:* G.C. Amstutz, A. El Goresy, G. Frenzel, C. Kluth, G. Moh, A. Wauschkuhn, and R.A. Zimmermann, Eds., Ore Genesis: The State of the Art. Springer-Verlag, New York, Heidelberg, Berlin, p. 108-117.

Schwarz, H.P., and S.W. Burnie. 1973. Influence of sedimentary environment on sulfur isotope ratios in clastic rocks: a review: Min. Depos., v. 8, p. 264-277.

Schwertmann, U. 1959. Über die Synthese definierter Eisenoxyde unter verschiedenen Bedingungen: Zeits. anorg. allg. Chem., v. 298, p. 337-348.

Schwertmann, U., and H. Thalmann. 1976. The influence of Fe^{2+}, Si, and pH on the formation of lepidocrocite and ferrihydrite during oxidation of aqueous $FeCl_2$ solutions: Clay Min., v. 11, p. 189-200.

Seyfried, W.E., and J.L. Bischoff. 1977. Hydrothermal transport of heavy metals by seawater: the role of seawater/basalt ratio: Earth Plan. Sci. Let., v. 34, p. 71-77.

Seyfried, W.E., and J.L. Bischoff. 1979. Low temperature basalt alteration by seawater: an experimental study at 70°C and 150°C: Geochim. Cosmochim. Acta, v. 43, p. 1937-1947.

Seyfried, W.E., and J.L. Bischoff. 1981. Experimental seawater-basalt interaction at 300°C, 500 bars; chemical exchange, secondary mineral formation and implications for the transport of heavy metals: Geochim. Cosmochim. Acta, v. 45, p. 135-147.

Shanks, W.C., and J.L. Bischoff. 1977. Ore transport and deposition in the Red Sea geothermal system: a geochemical model: Geochim. Cosmochim. Acta, v. 41, p. 1507-1519.

Shanks, W.C., and J.L. Bischoff. 1980. Geochemistry, sulfur isotope composition, and accumulation rates of Red Sea geothermal deposits: Econ. Geol., v. 75, p. 445-459.

Sheldon, R.P. 1970. Sedimentation of iron-rich rocks of Llandovery Age (Lower Silurian) in the southern Appalachian basin. *In:* W.B.N. Berry and A.J. Boucot, Eds., Correlation of the North American Silurian Rocks. Geol. Soc. Amer., Spec. Paper 102, p. 107-112.

Sherwani, J.K. 1973. Computer Simulation of Ground Water Aquifers of the Coastal Plain of North Carolina. University of North Carolina, Water Resources Inst., Report No. 75, 126 pp.

Shockey, P.N., A.R. Renfro, and R.J. Peterson. 1974. Copper-silver solution fronts at Paoli, Oklahoma: Econ. Geol., v. 69, p. 266-268.

Siehl, J.A., and J. Thien. 1978. Geochemisches Trends in der Minette (Jura, Luxembourg/ Lothringen): Geol. Rundschau, v. 67, p. 1052-1077.

Sillitoe, R.H. 1980. Are porphyry copper and Kuroko-type massive sulfide deposits incompatible?: Geology, v. 8, p. 11-14.

Sillitoe, R.H., and A.H. Clark. 1969. Copper and copper-iron sulfides as the initial products of supergene oxidation, Copiapo mining district, northern Chile: Amer. Min., v. 54, p. 1684-1710.

Simpson, P.R., and J.F.W. Bowles. 1977. Uranium mineralization in the Witwatersrand and Dominion Reef systems: Phil. Trans. Roy. Soc. London, v. A286, p. 527-548.

Simpson, P.R., and J.F.W. Bowles. 1981. Detrital pyrite and uraninite: are they evidence for a reducing atmosphere? *In:* F.C. Armstrong, Ed., Genesis of Uranium- and Gold-Bearing Precambrian Quartz-Pebble Conglomerates. U.S. Geol. Survey, Prof. Paper 1161, p. S1-S12.

Sinclair, I.G. 1966. Estimation of the thickness of limestone required to produce the Jamaican bauxites by the residual process: Jour. Geol. Soc. Jamaica, v. 8, p. 24-31.

Sinclair, I.G.L. 1967. Bauxite genesis in Jamaica: new evidence from trace element distribution: Econ. Geol., v. 62, p. 482-486.

Sivaprakash, C. 1980. Mineralogy of manganese deposits of Kodura and Garbham, Andhra Pradesh, India: Econ. Geol., v. 75, p. 1083-1104.

Skinner, B.J. 1958. The geology and metamorphism of the Nairne pyritic formation, a sedimentary sulfide deposit in South Australia: Econ. Geol., v. 53, p. 546-562.

Skinner, B.J., D.E. White, H.J. Rose, and R.E. Mays. 1967. Sulfides associated with the Salton Sea geothermal brine: Econ. Geol., v. 62, p. 316-330.

Sklijarov, R.J. 1978. The global law-governed nature of the development of bauxite and alunite deposits. *In:* S.S. Augustithis, Ed., 4th International Congress for the Study of Bauxites, Alumina and Aluminum. Greece, Natl. Tech. University Athens, p. 797-805.

Slee, K.J. 1980. Geology and origin of the Groote Eylandt manganese oxide deposits, Australia. *In:* I.M. Varentsov and Gy. Grasselly, Eds., Geology and Geochemistry of Manganese. E. Schweizerbart'sche Verlagsbuchhandlung, Stuttgart, v. 2, p. 125-148.

Smith, B.H. 1977. Some aspects of the use of geochemistry in the search for nickel sulphides in lateritic terrain in Western Australia: Jour. Geochem. Exploration, v. 8, p. 259-281.

Smith, D.I. 1970. The residual hypothesis for the formation of Jamaican bauxite—a consideration of the rate of limestone erosion: Geol. Soc. Jamaica Jour., v. 11, p. 3-12.

Smith, G.E. 1976. Sabkha and tidal-flat facies control of stratiform copper deposits in North Texas: Okalahoma Geol. Survey, Circular 77, p. 25-39.

Smith, J.W., and N.J.W. Croxford. 1973. Sulphur isotope ratios in the McArthur lead-zinc-silver deposit: Nature Phys. Sci., v. 245, p. 10-12.

Smith, J.W., and N.J.W. Croxford. 1975. An isotopic investigation of the environment of deposition of the McArthur mineralization: Min. Depos., v. 10, p. 269-276.

Smith, N.D., and W.E.L. Minter. 1980. Sedimentologic controls of gold and uranium in two Witwatersrand paleoplacers: Econ. Geol., v. 75, p. 1-26.

Smythe, J.A., and K.C. Dunham. 1947. Ankerites and chalybites from the northern Pennine orefield and the north-east coalfield: Jour. Min. Soc., v. 28, p. 53-73.

Soloman, M., and J.L. Walshe. 1979. The formation of massive sulfide deposits on the sea floor: Econ. Geol., v. 74, p. 797-813.

Sorem, R.K., and E.N. Cameron. 1960. Manganese oxides and associated minerals in the Nsuta manganese deposits, West Africa: Econ. Geol., v. 55, p. 278-310.

Sorem, R.K., and R.H. Fewkes. 1979. Manganese Nodules. Research Data and Methods of Investigation. IFI/Plenum, New York, 723 pp.

Sorem, R.K., and A.R. Foster. 1972. Internal structure of manganese nodules and implications in beneficiation. In: D.R. Horn, Ed., Ferromanganese Deposits on the Ocean Floor. Lamont-Doherty Geol. Observatory, Palisades, New York, p. 167-182.

Sorem, R.K., and Gunn, D.W. 1967. Mineralogy of manganese deposits, Olympic Peninsula, Washington: Econ. Geol., v. 62, p. 22-56.

Spooner, E.T.C., and C.J. Bray. 1977. Hydrothermal fluids of seawater salinity in ophiolitic sulphide ore deposits in Cyprus: Nature, v. 266, p. 808-812.

Spooner, E.T.C., and W.S. Fyfe. 1973. Sub-sea floor metamorphism, heat and mass transfer: Contr. Min. Petrology, v. 42, p. 287-304.

Spooner, E.T.C., H.J. Chapman, and J.D. Smewing. 1977. Strontium isotope contamination and oxidation during ocean floor hydrothermal metamorphism of the ophiolitic rocks of the Troodos Massif, Cyprus: Geochim. Cosmochim. Acta, v. 41, p. 873-890.

Springer, G. 1974. Compositional and structural variations in garnierites: Canadian Min., v. 12, p. 381-388.

Stanton, R.L. 1976. Petrochemical studies of the ore environment at Broken Hill, New South Wales: Trans. Inst. Mining Metal., v. 85, p. B33-B47, B118-B131, B132-B141.

Strakhov, N.M. 1967. Principles of Lithogenesis. Consultants Bureau, New York, v. 1, 245 pp.

Strakhov, N.M., I.M. Varentsov, V.V. Kalinenko, E.S. Tikhomirova, and L.E. Shterenberg. 1970. The mechanism of manganese ore formation processes (Oligocene ores in the southern part of the USSR). *In:* D.G. Sapoznikov, Ed., Manganese Deposits of the Soviet Union. Israel Program for Scientific Translation, Jerusalem, p. 34-57.

Strong, D.F. 1973. Discussion of paper by Constantinou and Govett: Trans. Inst. Mining Metal., v. 82, p. B74.

Strong, D.F. 1977. Volcanic regimes of the Newfoundland Appalachians. *In:* W.R.A. Baragar, L.C. Coleman, and J.M. Hall, Eds., Volcanic Regimes in Canada. Geol. Assoc. Canada, Spec. Paper 16, p. 61-90.

Stuckless, J.S., C.M. Bunker, C.A. Bush, W.P. Doering, and J.H. Scott. 1977. Geochemical and petrological studies of a uraniferous granite from the Granite Mountains, Wyoming: Jour. Research U.S. Geol. Survey, v. 5, p. 61-81.

Stumm, W., and J.J. Morgan. 1970. Aquatic Chemistry. Wiley Interscience, New York, 583 pp.

Styrt, M.M., A.J. Brackmann, H.D. Holland, B.C. Clark, V. Pisutha-Arnold, C.S. Eldridge, and H. Ohmoto. 1981. The mineralogy and the isotopic composition of sulfur in hydrothermal sulfide/sulfate deposits on the East Pacific Rise, 21°N latitude: Earth Plan. Sci. Let., v. 53, p. 382-390.

Suess, E. 1979. Mineral phases formed in anoxic sediments by microbial decomposition of organic matter: Geochim. Cosmochim. Acta, v. 43, p. 339-352.

Sverjensky, D.A. 1981. The origin of a Mississippi Valley-type deposit in the Viburnum Trend, Southeast Missouri: Econ. Geol., v. 76, p. 1848-1872.

Swanson, E.A., D.F. Strong, and J.G. Thurlow. 1981. The Buchans Orebodies: Fifty Years of Geology and Mining. Geol. Assoc. Canada, Spec. Paper no. 22, 450 pp.

Symons, D.T.A. 1967. A paleomagnetic study of concentrating iron ores from northern Michigan: Econ. Geol., v. 62, p. 118-137.

Talbot, M.R. 1974. Ironstones in the Upper Oxfordian of southern England: Sedimentology, v. 21, p. 433-450.

Tardy, Y., and R.M. Garrels. 1974. A method of estimating the Gibbs energies of formation of layer silicates: Geochim. Cosmochim. Acta, v. 38, p. 1101-1116.

Tardy, Y., and R.M. Garrels. 1976. Prediction of Gibbs energies of formation—I. Relationships among Gibbs energies of formation of hydroxides, oxides and aqueous ions: Geochim. Cosmochim. Acta, v. 40, p. 1051-1056.

Tatsumi, T., Ed. 1970. Volcanism and Ore Genesis. University Tokyo Press, Tokyo, 448 pp.

Taylor, J.H. 1949. Petrology of the Northampton Sand Ironstone Formation. Great Britain Geol. Survey, Memoir, 111 pp.

Taylor, J.H. 1951. Sedimentation problems of the Northampton Sand Ironstone: Proc. Yorkshire Geol. Soc., v. 28, p. 74-84.

Taylor, J.H., W. Davies, and R.J.M. Dixie. 1952. The petrology of the British Mesozoic ironstones and its bearing on the problems of beneficiation: 19th Intl. Geol. Cong., Symposium sur les Gisements de Fer du Monde. v. 2, p. 453-466.

Taylor, S., and C.J. Andrew. 1978. Silvermines orebodies, County Tipperary, Ireland: Trans. Inst. Mining Metal., v. 87, p. B111-B124.

Thompson, G.R., and J. Hower. 1975. The mineralogy of glauconite: Clays Clay Min., v. 23, p. 289-300.

Thompson, R.I., and A. Panteleyev. 1976. Stratabound mineral deposits of the Canadian Cordillera. *In:* K.H. Wolfe, Ed., Handbook of Strata-Bound and Stratiform Ore Deposits. Elsevier, Amsterdam, v. 5, p. 37-108.

Thornber, M.R., and J.E. Wildman. 1979. Supergene alteration of sulphides, IV. Laboratory study of the weathering of nickel ores: Chem. Geol., v. 24, p. 97-110.

Thurlow, J.G., E.A. Swanson, and D.F. Strong. 1975. Geology and lithogeochemistry of the Buchans polymetallic sulfide deposit, Newfoundland: Econ. Geol., v. 70, p. 130-144.

Thurlow, J.G., and E.A. Swanson. 1981. Geology and ore deposits of the Buchans area central Newfoundland. *In:* E.A. Swanson, D.F. Strong, and J.G. Thurlow, Eds., The Buchans Orebodies: Fifty Years of Geology and Mining. Geol. Assoc. Canada, Spec. Paper no. 22, p. 113-142.

Timofeyeva, Z.V., L.D. Kuznetsova, and Ye.I. Dontsova. 1976. Oxygen isotopes and siderite formation: Geochem. Intl., No. 10, p. 101-112.

Titley, S.R., and C.L. Hicks, Eds. 1966. Geology of the Porphyry Copper Deposits. University of Arizona Press, Tucson, 287 pp.

Tourtelot, E.B., and J.D. Vine. 1976. Copper Deposits in Sedimentary and Volcanic Rocks. U.S. Geol. Survey, Prof. Paper 907-C, 34 pp.

Towe, K.M., and T.T. Moench. 1981. Electron-optical characterization of bacterial magnetite: Earth Plan. Sci. Let., v. 52, p. 213-220.

Trendall, A.F. 1968. Three great basins of Precambrian banded iron formation deposition: a systematic comparison: Geol. Soc. Amer. Bull., v. 79, p. 1527-1533.

Trendall, A.F. 1973a. Precambrian iron-formations of Australia: Econ. Geol., v. 68, p. 1023-1034.

Trendall, A.F. 1973b. Varve cycles in the Weeli Wolli Formation of the Precambrian Hamersley Group, Western Australia: Econ. Geol., v. 68, p. 1089-1097.

Trendall, A.F. 1973c. Iron-formations of the Hamersley Group of Western Australia: type examples of varved Precambrian evaporites. *In:* Genesis of Precambrian Iron and Manganese Deposits. UNESCO, Paris, p. 257-270.

Trendall, A.F., and J.G. Blockley. 1970. The iron formations of the Precambrian Hamersley Group, Western Australia: Geol. Survey Western Australia, Bull., v. 119, p. 3-346.

Troly, G., M. Esterle, B.G. Pelletier, and W. Reibell. 1979. Nickel deposits in New Caledonia—some factors influencing their formation. *In:* D.J.I. Evans, R.S. Shoemaker, and H. Veltman, Eds., International Laterite Symposium. Amer. Inst. Mining, Metal., and Petroleum Engineers, New York, p. 85-120.

Truswell, F., and K.A. Eriksson. 1973. Stromatolitic associations and their paleoenvironmental significance: a reappraisal of a lower Proterozoic locality from the northern Cape Province, South Africa: Sed. Geol., v. 10, p. 1-23.

Truswell, F., and K.A. Ericksson. 1975. A paleoenvironmental interpretation of the early Proterozoic Malmani dolomite from Zwartkops, South Africa: Precambrian Res., v. 2, p. 277-303.

Tuach, J., and M.J. Kennedy. 1978. The geologic setting of the Ming and other sulfide deposits, Consolidated Rambler Mines, Northeast Newfoundland: Econ. Geol., v. 73, p. 192-206.

Turner, J.S., and L.B. Gustafson. 1978. The flow of hot saline solutions from vents in the sea floor—some implications for exhalative massive sulfide and other ore deposits: Econ. Geol., v. 73, p. 1082-1100.

Upadhyay, H.D., and D.F. Strong. 1973. Geologic setting of the Betts Cove copper deposits, Newfoundland: an example of ophiolite sulfide mineralization: Econ. Geol., v. 68, p. 161-167.

Urabe, T., and T. Sato. 1978. Kuroko deposits of the Kosaka Mine, northeast Honshu, Japan—Products of submarine hot springs on Miocene sea floor: Econ. Geol., v. 73, p. 161-179.

U.S. Department of Energy. 1977. Preprints for First Eastern Gas Shales Symposium, Morgantown Energy Technology Center, Morgantown, WV, MERC/SP-77/5, 563 pp.

U.S. Department of Energy. 1978. Preprints for Second Eastern Gas Shales Symposium, Morgantown Energy Technology Center, Morgantown, WV, METC/SP-78/6, v. 1, 454 pp.; v. 2, 163 pp.

U.S. Department of Energy. 1979. Proceedings of Third Eastern Gas Shales Symposium, Morgantown Energy Technology Center, Morgantown, WV, METC/SP-79/6, 542 pp.

Usui, A. 1979. Nickel and copper accumulation as essential elements in 10Å manganite of deep-sea manganese nodules: Nature, v. 279, p. 411-413.

Uytenbogaardt, W., and E.A.J. Burke. 1971. Tables for Microscopic Identification of Ore Minerals. Elsevier, Amsterdam, 430 pp.

Valeton, I. 1964. Problems of boehmitic and diasporitic bauxites. In: G.C. Amstutz, Ed., Sedimentology and Ore Genesis. Elsevier, Amsterdam, p. 123-129.

Valeton, I. 1972. Bauxites. Elsevier, Amsterdam, 226 pp.

Valette, J.N. 1973. Distribution of certain trace elements in marine sediments surrounding Vulcano Island (Italy). In: G.C. Amstutz and A.J. Bernard, Eds., Ores in Sediments. Springer-Verlag, New York, Heidelberg, Berlin, p. 321-338.

Van Andel, Tj., and H. Postma. 1954. Recent sediments of the Gulf of Paria: Verhand. Koninklije Nederlandse Akad. Wetenschappen, v. 20, p. 1-244.

van Eden, J.G. 1974. Depositional and diagenetic environment related to sulfide mineralization, Mufulira, Zambia: Econ. Geol., v. 69, p. 59-79.

Van Houten, F.B., and D.P. Bhattacharyya. 1982. Phanerozoic oolitic ironstones—geologic record and facies: Ann. Rev. Earth Plan. Sci., v. 10, p. 441-458.

Van Houten, F.B., and R.M. Karasek. 1981. Sedimentologic framework of Late Devonian oolitic iron formation, Shatti Valley, west-central Libya: Jour. Sed. Petrology, v. 51, p. 415-427.

Varentsov, I.M. 1964. Sedimentary Manganese Ores. Elsevier, Amsterdam, 119 pp.

Varentsov, I.M., and V.P. Rakhmanov. 1977. Deposits of manganese. In: V.I. Smirnov, Ed., Ore Deposits of the USSR. Pittman, London, p. 114-178.

Varentsov, I.M., and V.P. Rakhmanov. 1980. Manganese deposits of the USSR (a review). *In:* I.M. Varentsov and Gy. Grasselly, Eds., Geology and Geochemistry of Manganese. E. Schweizerbart'sche Verlagsbuchhandlung, Stuttgart, v. 2, p. 319-392.

Veizer, J., and J. Hoefs. 1976. The nature of $^{18}O/^{16}O$ and $^{13}C/^{12}C$ secular trends in sedimentary carbonate rocks: Geochim. Cosmochim. Acta, v. 40, p. 1387-1395.

Veizer, J., W.T. Holser, and C.K. Wilgus. 1980. Correlation of $^{13}C/^{12}C$ and $^{34}S/^{32}S$ secular variations: Geochim. Cosmochim. Acta, v. 44, p. 579-587.

Viljoen, R.P., R. Sager, and M.J. Viljoen. 1980. Some thoughts on the origin and processes responsible for the concentration of gold in the early Precambrian of southern Africa: Min. Depos., v. 5, p. 164-180.

Vincienne, H. 1956. Observations géologiques sur quelques gîtes Marocains de manganèse syngenétique. *In:* J.G. Reyna, Ed., Symposium on Manganese. 20th Intl. Geol. Cong. Mexico, p. 249-268.

von Backstrom, J.W. 1981. The Dominion Reef Group, Western Transvaal, South Africa. *In:* F.C. Armstrong, Ed., Genesis of Uranium- and Gold-Bearing Precambrian Quartz-Pebble Conglomerates. U.S. Geol. Survey, Prof. Paper 1161, p. F1-F8.

Walker, R.G. 1978. A critical appraisal of Archean basin-craton complexes: Canadian Jour. Earth Sci., v. 15, p. 1213-1218.

Wall, J.R.D., E.B. Wolfenden, E.H. Beard, and T. Deans. 1962. Nordstrandite in soil from West Sarawak, Borneo: Nature, v. 196, p. 264-265.

Walton, A.W., W.E. Galloway, and C.D. Henry. 1981. Release of uranium from volcanic glass in sedimentary sequences: an analysis of two systems: Econ. Geol., v. 76, p. 69-88.

Warne, S. St.J. 1962. A field or laboratory staining scheme for the differentiation of the major carbonate minerals: Jour. Sed. Petrology, v. 32, p. 29-38.

Warren, G.C. 1972. Sulfur isotopes as a clue to the genetic geochemistry of a roll-type uranium deposit: Econ. Geol. v. 67, p. 759-767.

Watmuff, I.G. 1974. Supergene alteration of the Mt. Windarra nickel sulphide ore deposit, Western Australia: Min. Depos., v. 9, p. 199-221.

Weber, J.N., E.C. Williams, and M.L. Keith. 1964. Paleoenvironmental significance of carbon isotopic composition of siderite nodules in some shales of Pennsylvanian age: Jour. Sed. Petrology, v. 34, p. 814-818.

Wedepohl, K.H. 1964. Untersuchungen am Kupferschiefer in nordwest Deutschland: Ein Beitrag zur Deutung der Genese bituminöser Sedimente: Geochim. Cosmochim. Acta, v. 28, p. 305-364.

Wedepohl, K.H., Ed. 1969. Handbook of Geochemistry. Springer-Verlag, New York, Heidelberg, Berlin, various paging. Wedepohl, K.H. 1971a. Geochemistry, Holt, Rinehart and Winston, New York, 231 pp.

Wedepohl, K.H. 1971b. Kupferschiefer as a prototype of syngenetic sedimentary ore deposits: Soc. Mining Geol. Japan, v. 3, p. 268-273.

Wedepohl, K.H. 1980. Geochemical behavior of manganese. *In:* I.M. Varentsov and Gy. Grasselly, Eds., Geology and Geochemistry of Manganese. E. Schweizerbart'sche Verlagsbuchhandlung, Stuttgart, v. 1, p. 335-352.

Weinberg, R.M. 1973. The Petrology and Geochemistry of the Cambro-Ordovician Iron-stones of North Wales. Unpubl. PhD Dissertation, Oxford University, Oxford, 350 pp.

White, A.H. 1976. Genesis of low iron bauxite, northeastern Cape York, Queensland, Australia: Econ. Geol., v. 71, p. 1526-1532.

White, D.A. 1954. The structure and stratigraphy of the Mesabi Range, Minnesota: Minnesota Geol. Survey, Bull. 38, 92 pp.

White, D.E. 1968. Environments of generation of some base-metal ore deposits: Econ. Geol., v. 63, p. 301-335.

White, W.S. 1968. The native copper deposits of northern Michigan. In: J.E. Ridge, Ed., Ore Deposits of the United States. Amer. Inst. Mining, Metal., and Petroleum Engineers, v. 1, p. 308-325.

White W.S. 1971. A paleohydrologic model for mineralization of the White Pine copper deposit: Econ. Geol., v. 66, p. 1-13.

White, W.S., and J.C. Wright. 1966. Sulfide mineral zoning in the basal Nonesuch Shale, northern Michigan: Econ. Geol., v. 61, p. 1171-1190.

Whitehead, R.E.S., and W.D. Goodfellow. 1978. Geochemistry of volcanic rocks from the Tetagouche Group, Bathurst, New Brunswick, Canada: Canadian Jour. Earth Sci., v. 15, p. 207-219.

Willden, M.Y. 1980. Paleoenvironment of the autochthonous sedimentary rock sequence at Laisvall, Swedish Caledonides: Stockholm Contributions in Geology, v. 33, p. 1-100.

Williams, G.E. 1978. Discussion of "Paleolatitudes of glaciogenic upper Precambrian Rapitan Group": Geology, v. 6, p. 4.

Williams, N. 1978a. Studies of the base metal sulfide deposits at McArthur River, Northern Territory, Australia: I. The Cooley and Ridge deposits: Econ. Geol., v. 73, p. 1005-1035.

Williams, N. 1978b. Studies of the base metal sulfide deposits at McArthur River, Northern Territory, Australia: II. The sulfide-S and organic-C relationships of the concordant deposits and their significance: Econ. Geol., v. 73, p. 1036-1056. Discussion, v. 74, p. 1693-1702.

Williams, N., and D.M. Rye. 1974. Alternative interpretation of sulphur isotope ratios in the McArthur lead-zinc-silver deposit: Nature, v. 247, p. 535-537.

Wilson, J.L. 1975. Carbonate Facies in Geologic History, Springer-Verlag, New York, Heidelberg, Berlin, 477 pp.

Wolery, T.J., and N.D. Sleep. 1976. Hydrothermal circulation and geochemical flux at mid-ocean ridges: Jour. Geol., v. 84, p. 249-275.

Wolery, T.J., and N.D. Sleep. in press. Interaction of exogenic and endogenic cycles. In: B. Gregor and F.T. Mackenzie, Eds., Chemical Cycles in the Evolution of the Earth. John Wiley & Sons, New York.

Woodward, L.A., W.H. Kaufman, O.L. Schumacher, and L.W. Talbott. 1974. Strata-bound copper deposits in Triassic sandstone of Sierra Nacimento, New Mexico: Econ. Geol., v. 69, p. 108-120.

Yeats, P.A., B. Sundby and J.M. Bewers. 1979. Manganese recycling in coastal waters: Marine Chem., v. 8, p. 43-55.

Young, G. M. 1976. Iron-formation and glaciogenic rocks of the Rapitan Group, Northwest Territories, Canada: Precambrian Res., v. 3, p. 137-158.

Yu, H.-S. 1979. Three Aspects of Sandstone Diagenesis: Compaction and Cementation of Quartz Arenites and Chemical Changes in Graywackes. Unpubl. PhD Dissertation, University of Cincinnati, 135 pp.

Zans, V.A. 1959. Recent views on the origin of bauxite: Geonotes, v. 1, p. 123-132.

Zantop, H. 1978. Geologic setting and genesis of iron oxides and manganese oxides in the San Francisco manganese deposit, Jalisco, Mexico: Econ. Geol., v. 73, p. 1137-1149.

Zeissink, H.E. 1969. The mineralogy and geochemistry of a nickeliferous laterite profile (Greenvale, Queensland, Australia): Min. Depos., v. 4, p. 132-152.

Zielinski, R.A., D.A. Lindsey, and J.H. Rosholt. 1980. The distribution and mobility of uranium in glassy and zeolitized tuff, Keg Mountain area, Utah, U.S.A.: Chem. Geol., v. 29, p. 139-162.

Zimmerman, R.K., and S.E. Kesler. 1981. Fluid inclusion evidence for solution mixing, Sweetwater (Mississippi Valley-type) district, Tennessee: Econ. Geol., v. 76, p. 134-142.

Author Index

Subject Index

Index of Deposits
and Localities